A passion

FOR

protein

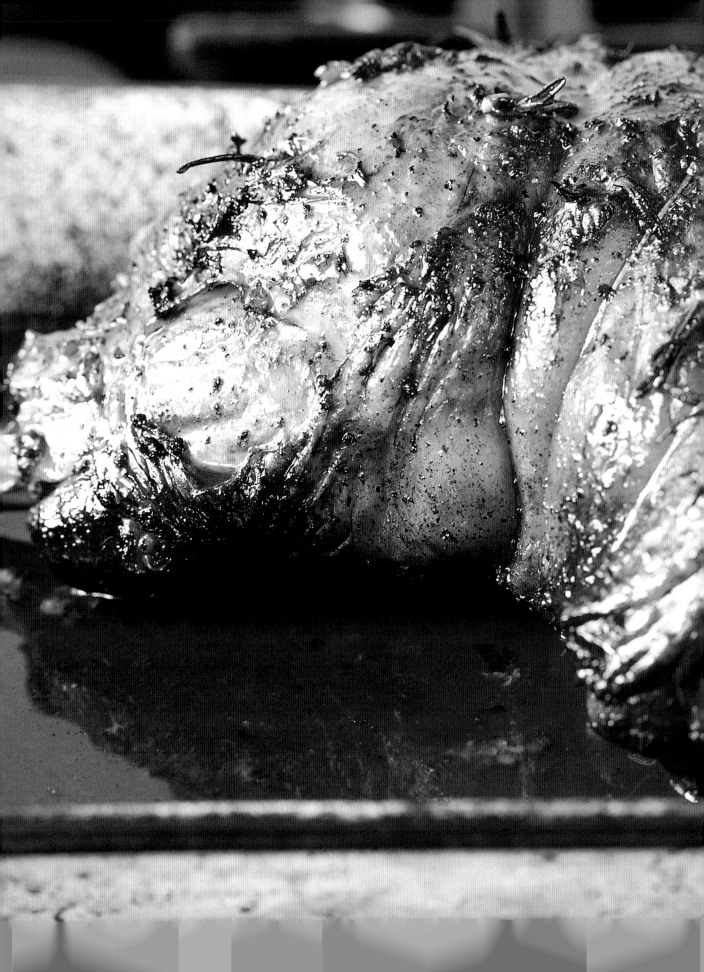

A passion FOR protein

High-protein, low-carbohydrate recipes for food lovers

HENRY HARRIS

PHOTOGRAPHY BY
JASON LOWE

For Georgia, Noah and Madeleine, my gorgeous children

First published in 2004 by
Quadrille Publishing Limited,
Alhambra House,
27-31 Charing Cross Road,
London WC2H OLS

Editorial Director: Jane O'Shea
Creative Director: Helen Lewis
Editor & Project Manager:
    Lewis Esson
Food Styling: Sunil Vijayakar
Art Director: Lawrence Morton
Production: Beverley Richardson

Cataloguing in Publication Data:
a catalogue record for this
book is available from the
British Library

ISBN 1 84400 102 4

Printed and bound in China

CONTENTS

INTRODUCTION

I think you can quickly see that this is no 'diet book' as such. There are no nutritional analyses of the recipes and using the book does not involve counting calories, etc. What I am attempting to do in these pages is produce a book for people just like me who really love good food but who want to lose some weight – something I recently achieved using very much the regime I outline here.

Credit where credit is due, it was my wife Denise who actually started the ball rolling. In August 2002, she began following the Atkins diet and by that Christmas she was buying new clothes for her much reduced size. Month by month I watched her become healthier, slimmer and happier, and with more energy – much needed for coping with our three lively children, aged three, eight and ten. Around April or May of 2003, it eventually permeated into my consciousness that if she could do it so could I. Until then I had steadfastly held the opinion that diets simply didn't fit in with my life as a chef, but what suddenly dawned on me was the realization that I had never EVER said no to ANY food or drink. Now I do… A healthy regime necessarily involves self control, and occasionally it does feel good to say no to your own urges.

Denise had followed the Atkins diet strictly but, having read the book she was using as her guide, I knew I just wouldn't be able to follow its diktats exactly without compromising what I really enjoy and what matters to me as a chef — and as a food lover. So I set about devising my own high-protein, low-carbohydrate regime, with recipes that are specifically designed for my own needs — and that of anyone who loves good food.

What had previously put me off any form of dieting programme was the fact that they always seemed to involve lowest common denominator dishes, poor-quality and canned food, artificial sweeteners and generally ersatz and dull stuff. Instead, I knew that for it to work for me the food had to be exciting and satisfying, and not that different from previous eating habits. I was also determined that my programme would be demonstrably healthy. So, despite the strictures of it being low-carb, it would have to include a wide range of vegetables (alas not fruits, because of their usually high sugar content), only eschewing starchy vegetables like potatoes, and grains and their products like rice, bread, pasta, etc.

However, just to kick-start things, I did begin with a week's sort of elimination diet, cutting out virtually all carbs – and as a consequence did suffer a continual stinking headache (an acknowledged short-term side effect of both elimination diets and low-carb diets). Later I reintroduced small amounts of carbs. I lost weight right from the start, but after about two weeks the weight loss suddenly increased dramatically. Within four months I had lost two stones (12.5kg) and I did feel much healthier and with renewed energy, while still enjoying almost all of the foods I love. I weighed myself a few days ago and I have lost 2½–3 stones (16–19kg) since I started on my new regime.

When I think about it, when I was a child in the late 60s – before everyone got obsessed with fat and cholesterol – whenever anyone wanted to diet, sugar, bread and potatoes were the first things to go. It is also becoming abundantly clear that it is the unprecedented amounts of added fat and refined sugar in modern convenience foods that are making our modern diet so problematic.

As well as having lost weight on my regime, I also have healthier skin, I sleep better – and need less sleep – and my energy levels are better. Before, if I had pasta I would digest it quickly and feel high for a short time and then feel lethargic. This sort of thing just doesn't happen now. Immediately after a meal I want to go and do things, and suffer little or no post-meal sluggishness.

You do recognize fairly quickly that the body works very differently on a high-protein, low-carb diet. As I no longer get carb rushes (proteins take longer to digest), I find it easier to go for longer without food. What I have learned about my body is much more

about what it really needs in the way of fuel. I now find that there are times when I don't need food and don't just eat it automatically because it is in front of me or because it is a certain time of day. I also drink more water, and not because I am thirsty, but simply because I want to drink it, and I also feel no need to snack with the water or any other drink.

Recently I was quite ill with a bad bout of the 'flu, which was followed by an upset stomach. All I wanted were mugs of Bovril and slices of comforting (but previously forbidden) brown toast and Marmite, fortunately moving on later to eggs and then steak. (My mother always says that when you are run down or recovering from an illness you should eat steak – 'It'll put you back on the map!') I could feel that my body needed these and when you are sick you need to listen to your body.

The main thing to remember about whenever your diet gets disturbed, be it by bouts of illness or the inevitable occasional giving in and bingeing, is just to get back on the straight and narrow as soon as you can. Don't beat yourself up and don't think you can't go on with it – the odd lapse is fairly inevitable, but more and more you'll find you don't miss the things you once thought you couldn't live without.

Lots of variety and a certain degree of flexibility are necessary to avoid getting bored, and you must build in the occasional indulgence so you don't feel like a slave to the regime. For example, I haven't eaten much cheese for a few days so I am really looking forward to my Baked Cooleney tonight – or I might even break out and have a pizza.

One of the most astonishing aspects of my weight experience is that periods of intense socializing and inevitable indulgence, like Christmas, don't take their usual toll. In fact, I weighed less on New Year's Eve 2003 than I had done at the beginning of December.

THE HIGH-PROTEIN, LOW-CARBOHYDRATE REGIME & HEALTH

There has been lots of media criticism recently about the possible negative effects of high-protein, low-carb diets and I felt I had to take this on board when I constructed my regime. Fortunately, my own love of vegetables took care of one of the main issues -– that of them not

8

supplying enough fibre. Getting enough veg also helps boost intake of vitamins, minerals and disease-fighting phytochemicals. Importantly, getting lots of one's protein from fish and poultry helps keep down the amount of saturated fat in the diet. My other passion, for cheese (yes, high in saturates, I know, just don't have too much), helps deal with the potential problem of possible calcium shortfall due to too much being excreted.

Nevertheless, while following this regime, I think it is very important to take high-potency vitamin/mineral supplements every day, as I do. If you are diabetic or have any potentially complicating condition, I would also recommend that you talk to your doctor before following the regime.

HIGH-PROTEIN NEEDN'T MEAN HIGH-COST Another negative often trotted out about high-protein eating is that it necessarily has to be expensive. Thankfully, this just isn't so, as long as you shop carefully and are flexible. Seasonality is everything – and not just with vegetables, but with meat, fish, cheeses, etc. Look for any gluts that are bringing down prices. Search the shops and markets for what is best and cheapest, and match that to one of my recipes; don't go shopping with fixed ideas in your head.

I try to favour street traders as their lower overheads generally mean better prices, and their usual lack of storage and rapid turnover usually guarantee greater freshness.

With fish, of course, a good fishmonger is everything. Sadly, they have become almost as endangered a species as some of the fish they can no longer sell. However, some better supermarkets now have excellent wet fish counters, manned by people who do know their stuff. We are lucky to have a Japanese fishmonger near our home who is incredibly helpful and a model of what a fishmonger should be like. His establishment and welcoming smile are as bright and fresh as his fish. There is also not a hint of stale fish smell. When you ask for fish to be scaled, gutted, trimmed or filleted, he unhesitatingly says 'Of course'; if yours doesn't, then walk away.

As much for reasons of favouring sustainable resources as budget, make a point of trying less popular basic white fish types, like pollack, whiting and gurnard. Also, ensure you get plenty of oily fish, like herring, mackerel and sardines for all those healthy omega 3 oils.

With meat, obviously go for the cheaper, tougher cuts with lots of connective tissue, like brisket and shins, where long slow cooking produces melting results with much more flavour.

Although poultry is incredibly good value for money at the moment, I would recommend that you be prepared to pay that little bit extra for corn-fed and/or free-range birds, as the cheaper birds tend to taste only of the poor stuff (often fish meal) on which they are fed. Don't go buying expensive breast fillets, but instead buy whole birds from which you can remove the breasts and then use the rest of the bird in so many ways to get the most from it for your money (see page 146).

Apart from calves' liver, most sorts of offal are usually available at bargain prices and lend themselves to all sorts of tasty dishes. For instance, lambs' liver and kidneys cooked gently in foaming butter make wonderful nutritious and satisfying meals – but do serve them pink, as overcooked they can be inedibly tough. Don't forget incredibly versatile chicken livers, which make excellent pâtés and one-pan sautés – or try chicken liver salad with sherry vinegar dressing.

FINDING SUBSTITUTES FOR CARBS You will see in the recipes that follow that I have developed a whole range of ways of replacing the carb elements in classic dishes: raw celery sticks instead of bread sticks, crumbled cooked bacon or pork scratchings in place of croutons in salads, chunks of Swedish herring as a nibble, using cooked large flat mushrooms as burger buns, finely shredded celeriac or courgette in place of pasta, lettuce and cabbage leaves as wraps. More often than not, however, it is not so much actually a matter of finding substitutes, but of reconstructing a dish or finding new combinations. Also bear in mind that a dish full of richness and flavour doesn't always necessarily need a potato-shaped accompaniment.

9

RECOGNIZING COVERT CARBS

For those wanting to be really exacting about keeping down carbohydrate intake, you do have to be aware that carbs are found in a much wider range of foods than you might at first think. Common sense will alert you to many such examples, such as items like fish in batter, breaded veal or chicken and white sauce, which are obviously high in carbs, and generally it is fairly easy to identify most hidden carbs, like the sugars in sweet veg such as carrots, tomatoes, peas, sweetcorn, parsnips and beetroots, but this can be a really tricky area. For instance, many people on such a regime lean heavily on nuts as high-protein snacks, but you have to be very careful as many types of nut, notably the cashew, are actually also high in carbs.

A treacherous area is low-fat food: things like semi-skimmed milk have a higher percentage of carbs than full-fat, as more milk has to be used to produce a given quantity of the low-fat version. The same holds true for items like cottage cheese and yoghurt. For all these reasons, I recommend anyone wishing to be fastidious about carb intake to arm themselves with a book of the nutritional values of common foods.

SOME ADVICE ON CONVENIENCE FOODS

The saving grace of ready-prepared foods is that they now all have to have complete nutritional breakdowns on the packs. Unfortunately, these usually do reveal too high a carb content, so you have to give up most of your favourites, but manufacturers and supermarkets are catching on to the vogue for low-carb diets and producing convenience food to fit the needs of those on this sort of regime. Don't forget really useful (and inexpensive) low-carb storecupboard standbys like canned tomatoes, anchovies, sardines and tuna.

DEALING WITH ALCOHOL

I do drink regularly, but I don't drink to excess, and will often go for days without alcohol. In the past I was always being told that in order to lose weight I would have to give up alcohol. However, I was able to lose my 2 stones and still drink, although I did stop having beer and have never been a fan of sweet liqueurs. Spirits are virtually carb-free so, my favourite vodka martini is totally acceptable, if not to be encouraged. Wines do have carbs, but if you go for dry not medium and certainly not dessert wines, you can stay on course. Ultra-brut champagne, with no dosage, from Laurent Perrier is ultra-dry, so is the low-carb dieter's best friend. Port and brandy are allowable occasionally for medicinal purposes. As with everything you must be responsible with your drinking – but I am not going to set out to dictate what you should do on this matter... it's up to you.

Certainly, if I feel I have put on a bit of weight, I do stop drinking completely and the weight goes. However, I do really enjoy a drink and just think of all those cancer-combating antioxidants in red wine...

THE PSYCHOLOGY OF STICKING TO A NEW REGIME

As I mentioned earlier, one of the most important pieces of advice is not to punish yourself when you fall from grace. If you brood over it, it will probably just make you more resentful of the strictures of the regime. Just start back as you were, and you'll find such lapses get less and less frequent, and the regime a more deep-seated aspect of your life.

An important point to remember is the paramount importance of breakfast. Food is our fuel and we can't start a day without it, however sedentary our occupation. We may think we simply haven't got the time for breakfast, but it is asking for trouble to try to live without it – you will end up having a sandwich, packet of crisps or chocolate bar in the middle of the morning. On the other hand, you can actually fry an egg in less time than it takes to brew a mug of tea.

For much the same reasons, for when those real hunger pangs strike it is also essential to have to hand a good array of the right type of snacks, be they squares of dark chocolate (at least 70% cocoa solids), chunks of cheese, olives or hard-boiled eggs.

Try not to eat too late in the day – no later than 3 hours before you plan to go to bed – as the later you eat the less weight you'll lose, because food then gets turned into a fat store. At weekends, try to have a big lunch and then a snack dinner in the evening.

THE IMPORTANCE OF EXERCISE One of the universal truths about losing weight is that you have to burn up more calories than you consume, so regular exercise is an essential element of this regime – as it is, or should be, on any weight-loss programme (or ongoing regime, such as life itself). Also, if you don't exercise you will lose lean tissue (muscle) as well as fat, which you really don't want to do for lots of reasons. Of course, this means that you won't lose as much weight, as muscle weighs more than fat, but you will look and feel better. Regular exercise will increase stamina and you'll look more healthy rather than gaunt.

To get the appropriate exercise needn't involve joining a gym, but simply, say, having lots of long brisk walks. However, the best results are obtained by a combination of aerobic and weight-bearing exercise; do the weight training (it could be done with dumb-bells at home or even just be some heavy gardening, say) every other day and the aerobic (running or brisk walking) on the days in between.

On this diet, I know I couldn't run a marathon (runners fuel up for such an event with huge platters of pasta), but I don't want to do that and I now know exactly what my body is capable of on it.

KIDS AND DIETING My kids eat well and eat proper fresh food most of the time. They eat the food Denise and I eat (and have tried all the recipes in book), but they do have rice and pasta plus lots of fruit as they need the energy. Interestingly, as a by-product of thinking more about what we eat, we have put even more thought into what the kids eat. Our aim is to make them love good fresh food so much they will never fall prey to the problem of the addictive combination of fat and sugar in fast food causing so much child obesity.

AND FINALLY Low-carb diets have certainly proved more efficient than others in reducing weight initially, but once you go back to previous eating habits you will certainly put it back on. You need to develop a maintenance regime with which you are happy to alter your diet for good.

This book isn't really about just losing weight... it is about establishing and maintaining a body weight with which you are comfortable and you feel happy about. It is also, more than anything, about eating well. If you try to diet and eat badly, you will feel bad. If you adopt an overweaningly austere regime... of course you will fail... and you will go right back to your old ways... stoking yourself with chips, other fatty foods and sweets.

I offer all the following recipes with considerable humility, as I am no dieting expert and have no formal training in nutrition. I simply know what worked for me and hope it will be of help to you. This is, in essence, a collection of recipes that gave me pleasure to create, and I hope it gives you the same fun in the kitchen that it gave me, and that you and your family and friends will really enjoy the food.

Henry Harris
London, January 2004

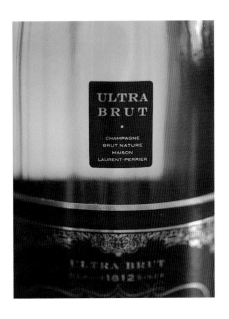

(label reads: ULTRA BRUT • CHAMPAGNE BRUT NATURE MAISON LAURENT-PERRIER)

BREAKFAST

The injunction to 'breakfast like a king and dine like a pauper' was never so true as it is on this sort of regime. Skip breakfast and you'll almost certainly end up giving in to some disastrous high-carb snack halfway through the morning. Unfortunately, there are just no easy substitutes for toast and breakfast cereals — instead, you have to make major adjustments to your thinking.

Apart from the traditional British cooked breakfast of bacon and egg, with optional sausage, which is a bit of a palaver if you are in a hurry, it is probably better if you think along the lines of ideas for brunch rather than traditional breakfast fare — although some other classic British breakfast favourites that have been rather forgotten in recent years can come back into their own — things like kippers and sautéed kidneys.

I usually have eggs in some form every day — boiled, fried scrambled or as an omelette.

I often keep a bowl of hard-boiled eggs in the fridge for days when I have no time at all in the morning or for snacks — mind you, as I said in my introduction, you can actually fry an egg in less time than it takes to brew tea.

Gone are the days of being implored to 'go to work on an egg', and although the egg has been rehabilitated in the eyes of the healthy eating lobby, there is still some anxiety about the wisdom of eating too many, due to their high cholesterol content. Rest assured, however, if you don't have an existing blood cholesterol problem, the body normally regulates its own levels, as long as the mechanisms are not subverted by too much saturated fat. So, make sure you are getting a high proportion of protein at other times of the day from fish and poultry.

If you really have no time to do anything but snatch something to eat en route, then there is always cheese, as well as cold cuts, chicken drumsticks and, of course, boiled eggs.

EGGS BENEDICT ON A POACHED SMOKED HADDOCK 'MUFFIN'

As you can see, there is actually no muffin here, its place is nicely taken by a piece of smoked haddock.

Serves 4

4 undyed smoked haddock fillets, each
  about 150g, skinned and trimmed
splash of white wine vinegar
4 eggs (the freshest you can find)
4 slices of cooked smoked ham

for the hollandaise sauce
3 tbsp white wine vinegar
250g unsalted butter
3 egg yolks

First make the hollandaise sauce: place the vinegar in a small pan, bring it to the boil and reduce its volume by half. Meanwhile, melt the butter in another small pan and ensure it is good and hot.

Pour the reduced vinegar into a food processor and add the egg yolks. Blitz the mixture for 30 seconds and then pour in the hot butter in a slow steady stream, just as you would for making mayonnaise. If the mixture becomes too thick, then add a splash of hot water. Season and set aside.

Place the smoked haddock fillets in a shallow pan and cover with cold water. Place the pan over a gentle heat, bring to a simmer and cook the fish for about 5 minutes or until it is just cooked, i.e. firm to the touch.

Meanwhile, heat another pan of water into which you have added a splash of vinegar and a good pinch of salt. Turn the heat down to the slowest of simmers and break in the whole eggs one by one. Let them sit at the bottom of the pan for 4 minutes so that the white sets and the yolk stays liquid.

With a fish slice, carefully lift each haddock fillet out of the water and place it on a plate. Top with a slice of ham. With a slotted spoon, lift each egg from the water and perch it on top of a slice of ham. With a piece of kitchen towel, mop up any cooking water. Finally spoon over the Hollandaise Sauce and serve.

14

SCRAMBLED EGGS WITH ANCHOVY

I have a love of Marmite on generously buttered toast that sadly had to be relinquished as part of my regime. How was I to get round it? After a little while, the simple – but, I think, slightly sad – method was to smear a little Marmite on a plate and then pour over the scrambled eggs. Absolutely delicious it was too. A friend from the rival Bovril camp suggested his preferred product and it worked with equal success. King of all additions was some best-quality salted anchovy fillet.

Serves 1

3 eggs
dash of Tabasco sauce
good knob of butter
splash of double cream
2 best-quality salted anchovy fillets, thinly sliced
  into slivers
1 tsp freshly chopped chives

Break the eggs into a bowl, add the Tabasco and a light seasoning of salt, and beat well.

Melt the butter in a heavy-based saucepan over a low heat and then pour in the egg mixture. Using a spatula, stir continuously, ensuring you keep lifting the setting egg from the bottom of the pan. Do this for several minutes or until the liquid egg is nearly all set. A very coarse porridge best describes the required texture.

Remove from the heat and stir in the cream. This halts the cooking process. Stir in the anchovy, spoon on to a plate and sprinkle over the chives.

16

BUTTERED KIPPERS

Provenance is vital on a kipper. A pair of small ones from Craster or large plump examples from Minola smoked foods are among my favourites. A good fishmonger or decent food hall should be able to provide you with some fine examples. My favourite cooking method for them is grilling, but I know that many of you will balk at the pervasive aroma that this produces. To prevent a dense oaky aroma from invading your kitchen, use the jug method, which works well but fails to give that depth of flavour that a good grilling will provide.

Serves 2

1 pair of large kippers or
  2 pairs of smaller ones
oil, for brushing (optional)
75g best-quality unsalted butter
2 lemon wedges

Preferred method: Preheat the grill. Place your kippers on a lightly oiled baking sheet or grill pan. Take a third of the butter and melt it in a mug in the microwave. Brush this over the kippers and then place them under the grill. Cook for 8–10 minutes, or until the fish is piping hot. (Don't place the kippers too high under the grill or they will scorch.) Remove them from the grill and transfer them to 2 plates. Divide the remaining butter in two and place it on the kippers. Season with a very generous milling of freshly ground black pepper and serve with the lemon wedges.

Less fragrant option (jug method without a jug): Fill a kettle with water and bring it to the boil. Place the kippers in a deep dish and pour over the boiling water. Leave for 10 minutes and then lift out carefully and place on plates. Dab dry with a piece of kitchen towel and proceed as above.

CHORIZO AND EGGS

A hearty spicy kick-start in the morning, this is more of a scrambled egg dish than an omelette. It seems all the more decadent with an accompanying Bloody Mary, especially as part of a weekend brunch.

Use an authentic Spanish chorizo, of the cooking sausage rather than salami type.

Should it be difficult to find, a young pecorino will be a reasonable substitute for the manchego cheese.

Serves 2

6 good fresh eggs
50g manchego cheese, grated
good splash of olive oil
125g chorizo (see above), skin removed and sliced

Break the eggs into a bowl, season with salt and pepper and beat well. Stir in the cheese.

Place a heavy non-stick pan over a medium heat and pour in the olive oil. Add the chorizo and fry until the colour of the sausage has taken on a darker hue and the meat is firm. Lift the sausage out of the pan and rest on a piece of kitchen towel.

Tip out the excess oil and return the pan to the heat. Pour in the egg mixture and stir slowly with a spatula, taking care to lift the setting curds of egg from the bottom of the pan. When the egg is still nicely creamy and wet, throw in the chorizo and fold it together.

Spoon on to 2 plates and consume immediately.

17

DEVILLED LAMBS' KIDNEYS

Spring and summer are the best times for lambs' kidneys. It is when their flavour is quite delicate and – when cooked gently – rather similar in texture to that of a nicely poached egg. Traditionally a devilled sauce is a dark stock-based sauce, but this creamed version requires only a little fresh stock and is easier to prepare. I approve of this old-fashioned dish as a start to the day, as it really is most satisfying. If it seems too stout for you in the morning, then pair it with a green salad to make a good supper dish.

Serves 2

knob of butter
6 lambs' kidneys, halved
1 tbsp chopped shallot
3 tbsp red wine vinegar
2 tbsp Worcestershire sauce

splash of brandy
100ml chicken stock
100ml whipping cream
1 tsp Dijon mustard
1 tsp grainy mustard

Melt the butter in a frying pan until foaming. Season the kidneys and place them, cut side down, in the butter. Cook over a medium heat for 2 minutes, turn them over and cook for a further 2 minutes. Lift them from the pan and keep warm.

   Throw the shallot into the pan and cook for a minute or two until softened. Add the vinegar and Worcestershire sauce, and boil it until the liquid has all but evaporated. Add the brandy, cook for 1 minute then add the stock and cream. Simmer and reduce the liquid until it has thickened a little. Adjust the seasoning, if necessary. It should have a gentle spiciness to it.

   Whisk in the two mustards and then return the kidneys to the sauce together with any juices, and give a quick and final boil before serving.

SNACKS AND S

This chapter is a rather picaresque assortment of snacky-type dishes, party food and nibbles, and one or two basics that will come in handy all over the place, like my basic vinaigrette. The actual snack dishes featured are far from an exhaustive selection, and what I present here should mainly serve as a guide to what is possible and help get you thinking more about the foods around you.

Snacks form another area in which one really needs to develop a whole new way of thinking. Simply by association (and dint of the implacable forces of global marketing), we have all become convinced that snacks need to be packed with sugar and fat. It doesn't really take much thought to see beyond this: cherry tomatoes, celery stalks, olives and handfuls of seeds and nuts make excellent snacks at any time. Nuts and seeds are also very healthy in

UNDRIES

that they are rich in the essential omega oils and other goodies, but do watch out as some nuts, like the cashew, are also very high in carbs. The vegetable crudités are also even more satisfying if served with one of the dips in this chapter.

Good high-protein snacks include chunks of cheese and dried beef, like jerky. These are, however, also high in saturated fat and so should only be used as an occasional treat and not made part of your regular daily regime. If you have to have something sweet (and, to be honest, cherry tomatoes can sometimes do it for me), then one square of good high-cocoa-solids dark chocolate makes a very satisfying snack at any time and contains relatively little sugar — only habit that makes you think you need more than one!

COD'S ROE AND SMOKED SALMON PÂTÉ/DIP

Taramasalata had to come off the menu as it is full of white breadcrumbs, but this alternative more than made up for it. I have developed a slight addiction for this, together with celery sticks which provide a great contrast in texture and make a wonderful dipping tool.

Makes about 600g

300g smoked cod's roe
100g smoked salmon pieces
1 garlic clove, crushed
generous pinch of cayenne pepper
equally generous milling of black pepper
juice of 1 small lemon
75g cream cheese
125ml olive oil

Pare the skin from the cod's roe and place the roe in a food processor together with all the other ingredients except for the olive oil. Blitz the mixture until it is smooth. With the machine still running, add the oil in a very slow steady stream as you would making mayonnaise. If the mixture looks as if it is about to split, then add a drop of boiling water from the kettle.

Check the seasoning as more lemon juice and pepper are sometimes needed. Transfer to a bowl and chill in the fridge for an hour before serving.

22

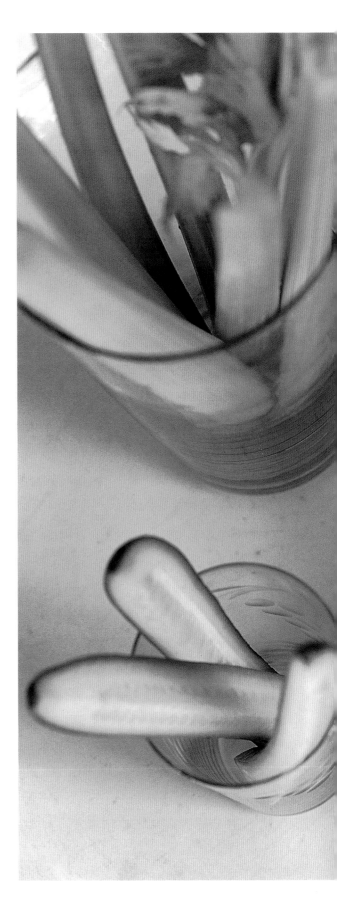

PRAWN COCKTAIL CANAPÉS

Recently I was asked to provide some retro canapés for a party. Prawn cocktail was asked for and instead of trying to make silly little bowls to serve them in I turned to that old retro favourite of pineapple chunks and Cheddar on sticks. It looked fab and has since become a favourite. It is quick and easy for a festive drinks party, particularly if you cheat and use ready-made mayonnaise. A note of caution from my mother; don't serve it in a room with a carpet.

makes 40 sticks

½ small white cabbage
1 cucumber, cut into small neat cubes
40 large peeled prawns

for the cocktail sauce
500g good-quality mayonnaise
2 tbsp tomato ketchup
Worcestershire sauce, to taste
Tabasco sauce, to taste
2 tbsp grated horseradish
1 tbsp Dijon mustard
2 tbsp Cognac (optional but good)
juice of ½ lemon

First make the cocktail sauce: empty the mayonnaise into a bowl and add the remaining ingredients with salt to taste. Stir well and add a little more of any of these flavouring ingredients should you feel it necessary.

Set the cabbage half on a serving platter. Thread a piece of cucumber on a wooden satay stick so it resides at the blunt end. Follow it with a prawn and a second piece of cucumber. Insert these sticks into the cabbage until there is room for no more.

Fill a couple of pots with the cocktail sauce and serve alongside the prawns; alternatively, flood the serving platter with the sauce.

AN AUTHENTIC CAESAR SALAD

It came as a bit of a shock to discover that Caesar salad was not originally made with anchovies in the dressing and, after trying it this way, I became converted. I know the English mustard wasn't originally there, but it gives the dressing a sharpness that you can't get from lemon juice alone.

Used as a dressing over shredded Cos lettuce, with more Parmesan and a few anchovy fillets, this will give you a wonderful – but crouton-less – Caesar salad. (If you really can't live without a crouton crunch, add some pork scratchings.)

Alternatively, drizzle the Parmesan dressing over cold broccoli and shredded poached chicken. I do feel obliged here to remind you of the inadvisability of serving raw eggs to children, the elderly, pregnant women and invalids.

Serves 4

2 small Cos lettuces, separated
8 canned anchovies in olive oil, well
  drained
chunk of Parmesan cheese

for the Caesar dressing
2 eggs
juice of 1 lemon
1 garlic clove, bashed
4 heaped tbsp freshly grated Parmesan
  cheese
½ tsp English mustard
250ml olive oil
Tabasco sauce, to taste
Worcestershire sauce, to taste

Bring a small pan of water to the boil, lower in the eggs and boil them for 1 minute precisely. Remove them from the water and set them aside for 2 minutes precisely.

Cut the tops off of the eggs and scoop out all of the contents into a liquidizer or food processor. Add the lemon juice, garlic, Parmesan and mustard. Run the liquidizer to blend it all together and then add the olive oil slowly, as you would making mayonnaise. As mentioned elsewhere in the book, have a kettleful of hot water to hand should you need to thin the dressing down a little. Finally season with Tabasco and Worcestershire sauces.

Separate the lettuce leaves and put into a large salad bowl, dot with the anchovies, pour over the dressing and toss well to coat. Shave over lots more Parmesan to serve.

SILKEN TOFU AND TAHINI DIP

This recipe originally had a little sugar in it – 1½ teaspoons of Demerera if you are tempted – but I find that the natural sweetness of the vegetables you dip into it provides all the seasoning required. Note that tofu does have a small carb content.

Serves 4–6

1 carton (290g) of soft silken tofu
150g mascarpone cheese
2 tbsp tahini
4 tbsp soy sauce, ideally Kikkoman, plus a little
   extra to taste if required
1 fennel bulb, cored
1 red pepper, deseeded
large bunch of spring onions

Remove the tofu from its carton, wrap it in some paper towel and microwave it for 2 minutes on the lowest or defrost setting. Remove from the microwave cooker and set aside to allow it to give up any residual heat.

When cooled, place it in a food processor with the mascarpone, tahini and soy sauce, and blend until smooth. Adjust the seasoning, if necessary; because of the natural blandness of the tofu and mascarpone, it may well need a little more soy sauce. Chill until needed.

Cut the fennel and red pepper into batons and trim the spring onions, leaving them whole. Serve the vegetables alongside the dip.

BASIC VINAIGRETTE

This is my basic 'no frills' dressing to be used whenever a vinaigrette is called for.
Using the soy sauce in place of salt gives an aromatic warmth that salt can't give. I don't often recommend a branded product but in this case I recommend you use Kikkoman soy sauce, as it is the most gentle of seasonings.

1 heaped tsp Dijon mustard
4 tbsp Kikkoman soy sauce
¼ tsp ground black pepper
2 tbsp red wine vinegar
130ml extra-virgin olive oil

Place the mustard, soy sauce, pepper and red wine vinegar in a bowl and whisk together. Then pour in the oil in a slow steady stream, while whisking all the time. A hand-held electric blender with its jug give the best results.

HOT STUFFED OLIVES

I ate these as a starter a few years ago in a pizzeria in Rome. They were so good that I started to make them at home. Usually one would pass them through milk and seasoned flour before frying, but to keep the carb content down I bake them in a warm oven, forgoing the flouring process.

Makes 36

2 slices of mortadella, each just under a centimetre in thickness (an Italian deli will do this)
36 large queen olives, stoned
olive oil
freshly ground black pepper

Preheat the oven to 180°C/gas 4. Cut the mortadella into cubes of a sufficient size so that each cube will nestle snugly into the olive's cavity. Place the stuffed olives in a baking dish, sprinkle over some olive oil and season with pepper only.

   Cover the dish with its lid or some foil and transfer to the oven. Bake for 20 minutes or until piping hot. Remove from the oven, transfer them to a serving dish and offer them around.

SPICED MACADAMIA NUTS

These are the ultimate high-protein, low-carb snack. The oil content of these nuts makes me sometimes wonder whether I have inadvertently eaten a ball of butter. Yes, they are rich. When purchased, they are usually well seasoned with salt but a subtle introduction of some spice and herb helps cut through the richness and makes them all the more tasty.

Makes 250g

250g salted macadamia nuts
sprinkling of cayenne pepper
4 allspice berries, ground to a powder
1 tsp finely chopped fresh dill

Place the nuts in a bowl, add the remaining ingredients and mix well. Serve!

29

MARINATED QUAILS' EGGS

Serve this as a moreish snack or part of a mezze-style selection of bits for a starter. Don't bother trying to make them in smaller quantities as you will run out too quickly!

Makes 24

24 quails' eggs
¼ tsp cumin seeds
¼ tsp coriander seeds
¼ tsp black peppercorns
1 tsp chopped parsley
½ tsp chopped thyme
finely grated zest of ½ lemon
½ tsp celery salt
½ tsp Maldon sea salt
drizzle of olive oil

Bring a pan of salted water to the boil. Gently lower in the eggs and cook for 4 minutes. Lift them from the pan, transfer them to a bowl and leave them under gently running cold water for 10 minutes to cool. When they are cool, shell the eggs and set aside in a bowl.

Place the cumin, coriander and peppercorns in a dry frying pan and dry-roast them for a minute to bring out their flavour. Using a mortar and pestle or small blender, grind them to a powder. Transfer the powder to a bowl and add the fresh herbs, lemon zest, celery salt and sea salt. Mix well.

Drizzle the oil over the eggs and gently move the eggs around until they are covered in the thinnest film of oil. Then scatter over your spice powder and once again move the eggs around until the powder is distributed as evenly as possible over the eggs. Cover them and store overnight in the fridge to allow the flavours to blend together.

Remove the eggs from the fridge at least half an hour before you are to consume them.

QUICK AND EA

I hope that the dishes I have featured in this section will prove to you that sticking to a regime like mine need not always involve you in complex and lengthy food preparation.

The recipes here require relatively little time spent in the kitchen and are all usually made from ingredients that cook quickly or are dishes that are easy to assemble and require little or no cooking, perhaps even just simple marination. If any lengthy cooking is involved, it is something that can happily be left unattended in the kitchen while you get on with other aspects of your life.

There are satisfying soups that are almost as quick to prepare as opening a can or a packet; fish recipes that require the lightest of cooking to produce dishes filled with flavour and all the natural goodness of the ingredients; lots of interesting ways with easily cooked vegetables, and fresh approaches to stalwart stand-bys like humble lamb cutlets and pork chops.

The message that I hope comes across here strongly is that convenience food need not mean expensive, additive-packed, ready-made stuff. If you have a well-stocked storecupboard, fridge and vegetable basket, then there is no reason why you can't readily conjure up meals that not only fit in perfectly with your diet, but are brimming with healthy nutrients – and are much more exciting and satisfying than the contents of any packet can ever be.

Moreover, not all the recipes in this section are for everyday fare, some are impressive enough for casual entertaining and one or two would easily grace any dinner party table. In fact, there is, I hope, something for every occasion when time is of the essence, as well as every taste.

PARSLEY SOUP

A good soup is the perfect way to start a meal and, for me, a not-too-thick cream soup seasoned with herbs, served with a glass of sherry on the side, is the closest to heaven that one will find in a bowl. On first glance this recipe may seem rather unexciting, but I assure you that it produces one of the most comforting and tasty soups I have ever had.

A good liquidizer is vital for blending soups, as most domestic food processors just aren't up to the job. At home I have blown up several cheaper models, I now use an American KitchenAid blender – its brute power purées the soup brilliantly and it doesn't expire midway through blending. You also need a fine sieve to pass the soup through; a cream soup should have a consistency barely thicker than single cream and have no solids present.

Serves 4

100g butter
5 large onions, thinly sliced
salt and freshly ground black pepper

1 litre fresh chicken stock
250ml double cream
large bunch of flat-leaved parsley
2 tbsp finely chopped shallot

Melt the butter in a large heavy pan and add the onions. Season, cover with a lid and cook over a very gentle heat, stirring regularly for about 20 minutes. When the onions are cooked to a slush-like consistency, add the stock, bring it to a gentle simmer and cook for a further 30 minutes, then add the cream (reserving a few spoonfuls for garnish if you like) and simmer for a further 10 minutes.

Meanwhile, take the parsley and twist off most of the stalks, then blanch the leaves for 20 seconds in a pan of boiling salted water.

Liquidize the soup in batches together with the parsley. Finally pass the soup through a sieve and return it to the pan.

To serve, give the soup a final seasoning of salt and pepper, reheat it carefully, ladle it into bowls and sprinkle over some of the shallot. If you like, trace the surface with a drizzle of cream.

HAKE BAKED WITH CHORIZO AND RED WINE

*This delicious Spanish-inspired recipe surprisingly cooks the fish in red wine –
you can use the leftover dregs from the previous night – but this in no way
overpowers the delicate flavour of the fish.*

Serves 4 as a main course

1 small onion, thinly sliced
4 slices of chorizo (of the salami-type
  variety and the spicier the better)

1 garlic clove, thinly sliced
4 skin-on fillets of hake or cod, each
  about 200g
200ml red wine
extra-virgin olive oil

Preheat the oven to 200°C/gas 6. Cut 4 sheets of kitchen foil each about 30x60cm,
then fold each over to make 4 double-layered 30cm squares and lay them out on a
work surface. Put some of the onion in the middle of each piece of foil, followed by a
slice of chorizo. Scatter a quarter of the garlic on top of the chorizo slices and finally
place a piece of fish on top. Crimp up the edges of the foil a little and pour some
wine into each parcel, then scatter with a seasoning of salt and pepper and a splash
of olive oil. Fold over the foil to seal each into a loose parcel, seal the edges well by
crimping them securely.

Place the parcels on a baking tray, transfer to the oven and bake for 15 minutes.

To serve, carefully open the parcels and lift the contents on to 4 plates and then
pour over the baking juices. You can also just serve the unopened parcels on plates
and let people open them themselves so they get the full benefit of the delicious
aroma as it rises. Serve with a spinach salad.

PORCINI SOUP

I always try to keep some dried morels and ceps/porcini in my store cupboard; morels for their intense smokiness and the ceps for their wonderful meatiness. Ceps are also ideal to use in dishes that are free of meat or fish as their strong flavour imparts real substance to any dish.

Serves 4

knob of butter
3 onions, finely chopped
500g flat mushrooms, thinly sliced
50g dried porcini
1 large garlic bulb, broken up into cloves and peeled
1 glass of white wine
1 litre fresh chicken stock
100g crème fraîche

Melt the butter in a large saucepan and add the onions, fresh and dried mushrooms, and the garlic. Turn down the heat, cover the pan and cook very gently for 40 minutes.

Add the white wine and simmer until reduced by half, then add the stock and bring to the boil. Simmer for a further 10 minutes, add the cream and remove from the heat.

Liquidize in small batches, pass through a sieve and return to a clean pan. Season with salt and pepper and serve.

ADDENDUM A big word of caution to those of you who are tempted to pick your own wild mushrooms, arming yourself with an illustrated guide could be fatal. If you are tempted to go picking, then go with someone who knows what they are looking for.

SMOKED HADDOCK MARINATED IN LIME AND RUM

Father Murray, a family friend and our former vicar, gave me this recipe for a remarkable and simple starter.

Serves 4 as a starter

600g fresh undyed smoked haddock
juice of 2 limes
4 tbsp dark rum
green tops of a bunch of spring onions

Go over the haddock with a pair of tweezers to remove any residual bones (or ask the fishmonger to do it for you). Then slice it at a shallow angle to give you plenty of thin slices, in the style of gravadlax. Arrange these in a single layer on 4 plates.

In a bowl, combine the lime juice with the rum, the most generous of seasonings of black pepper and a scant seasoning of sea salt. Spoon the liquid over the fish and leave for 5 minutes.

Finally, slice the spring onion tops and scatter them over the fish to serve.

TOMATOES BAKED WITH BASIL AND CRÈME FRAÎCHE

This is one of my most favourite recipes, I used to serve these tomatoes on toasted sourdough bread or – for a greater degree of refinement – toasted brioche. But sadly, now that bread is in the enemy camp, an alternative is needed. Enter a pork chop. Cook your pork chops as you usually do and garnish them with the tomatoes. The tomatoes are straightforward to prepare, but you may find that the sauce is too thick at the end; if so, then thin it down with a splash of hot water.

Serves 4 as a vegetable side dish

2 tbsp unsalted butter
4 large tomatoes
small bunch of basil, picked and coarsely chopped
200g crème fraîche

Melt half the butter in a cast-iron skillet or frying pan. Cut the tomatoes in half and season them with salt and pepper. Place them, cut side down, in the pan and cook over a very gentle heat for 10 minutes. The butter should foam and bubble and can take on the merest of light brown colouring, but no more.

   Turn the tomatoes over and add the crème fraîche. Leave on the heat for another 2 minutes. Adjust the seasoning, if necessary, and then stir in the basil. Serve immediately.

39

STEAMED MUSSELS WITH PASTIS AND FENNEL

This is a simple garlic-laden Provençal fast food number that gives an intriguingly different flavour to your mussels. Good mussels can be bought from your fishmonger ready-cleaned. Just give them a good rinse when you get home. Pick over them and discard any that stay open when tapped.

Serves 2 as a main course

3 tbsp olive oil
1 small dried red chilli, crushed
1 fennel bulb, halved and sliced as
   thinly as possible

2kg prepared mussels
4 garlic cloves, finely chopped
3 tbsp pastis, such as Ricard
2 tbsp chopped parsley (optional)

Put the olive oil and chilli in a large heavy pan that has a close-fitting lid and warm it over a gentle heat. Add the fennel to the pan and cook it gently until it has given up its crispness and turned soft and vaguely translucent, 10–15 minutes.

Add the garlic, cook for 20 seconds and add the mussels and pastis. Give everything a good working over with a large spoon to distribute the fennel mixture at the bottom of the pan amongst the mussels and then clamp on the lid. Cook for about 5 minutes over a high heat. You will need to check the mussels regularly. The power of your stove will determine the time it takes for the mussels to open. When the majority of the mussels are open, ladle them into two large bowls and spoon over the liquor and scatter the parsley if you are using it.

Consume, discarding those that have refused to open.

GRILLED LAMB CUTLETS WITH TAHINI DRESSING

I love this dressing/sauce/dip as it is so lively in flavour and so versatile. Middle-Eastern in origin, it is best known in a similar form as a sauce for falafel. I often have a pot in the fridge and serve it alongside grilled chicken or slow-roast shoulder of lamb. It is divine over grilled aubergine for a vegetarian meal and it makes a great dip for crudités.

The best chops for this are the cutlets cut from the saddle of lamb. Looking like a miniature T-bone steak, they are truly succulent when grilled, especially when barbecued over charcoal.

Serves 4 as a main course

12 lamb cutlets (see above)
freshly ground black pepper
olive oil, for brushing

for the tahini dressing
3 tbsp tahini
juice of 1 large lemon
1 garlic clove, finely grated or puréed
4 tbsp Greek-style yoghurt
splash of Tabasco sauce
100ml extra-virgin olive oil
splash of hot water

Put the tahini into a small bowl and stir in the lemon juice and garlic. When well blended, stir in the yoghurt, then stir in the Tabasco and olive oil with a generous seasoning of salt. Don't be shy with the salt, as the yoghurt seems to absorb excesses. Add a splash of hot water to make the dressing pourable.

Preheat a ridged cast-iron grill to a medium heat. Season the cutlets with salt and pepper and then brush them lightly on both sides with some olive oil. Grill them for about 8 minutes on each side to achieve a nice brown caramelization on the outside and a rosy-pink interior.

Set aside to rest in a warm place for at least 5 minutes and then serve with some of the tahini dressing drizzled over it.

WILTED GREENS AND PROSCIUTTO

Apart from the greens, these ingredients are traditional accompaniments to asparagus and if you want to 'posh' it up a bit then use asparagus as we have done here; however, the dish is just as nice without.

Serves 2 as a main course

4 slices of prosciutto (Parma ham)
bunch of slender young asparagus
   (optional)
good glug of Basic Vinaigrette (page 28)
2 shallots, finely chopped

bag of mixed salad (100–150g),
   containing a goodly proportion of
   rocket and watercress
chunk of Parmesan cheese

Cut each piece of ham in half and lay out 4 pieces of ham on each plate. If using asparagus, cook it in lightly salted boiling water for 3–4 minutes, rinse in cold water to stop the cooking, drain and then pat dry.

Put the vinaigrette into a wide saucepan together with the shallot and asparagus, and place it on a medium heat. Stir it around until the vinaigrette starts to splutter. Without hesitation, throw in salad and, using a pair of kitchen tongs, move the greens about to bathe them in the dressing. The leaves will start to wilt.

Immediately remove from the heat and, using the tongs, put a little pile of the leaves on each piece of ham. Roll up the ham slices like little cannelloni and spoon over some of the vinaigrette.

Finally, using a vegetable peeler, shave over a generous amount of Parmesan.

WHOLE ARTICHOKE VINAIGRETTE

This is summer food to linger over; it is simple to cook and it takes time to consume. Even for four you will need a very large pan. There is a common opinion that the cooking water of the artichoke should have lemon juice added to prevent the discoloration of the artichoke, but I find it masks the flavour and so leave it out and put up with some discoloration.

Serves 4 as a starter

4 large fresh artichokes
2 tsp Dijon mustard
3 tbsp red wine vinegar
200ml whipping cream

Bring a very large pan of salted water to the boil. When it is boiling, cut off and discard the stalks from the artichokes and introduce them into the boiling water. Cover with a lid and boil gently for 15–20 minutes. They are cooked when you can pull a leaf from the centre of the artichoke without much effort. Though do take care not to burn yourself. If the artichokes are particularly buoyant, then instead of the lid just rest a small pan on top of the artichokes to keep them submerged.

While the artichokes are cooking, place the mustard and vinegar in a bowl and stir together, along with a good seasoning of salt and pepper. Stir in the cream, adjust the seasoning if required and set this dressing to one side. Within 10 minutes the acidity of the vinegar will have thickened the cream a little.

When the artichokes are cooked, lift them from the pan and leave on a plate for 10 minutes to cool a little. They are best eaten lukewarm.

Give each person a plated artichoke and a ramekin of the simple cream dressing to dip it into. If you haven't tried an artichoke before, they can look a little intimidating but aren't at all. Start pulling the leaves off from the outside. At the base of each leaf is a small fleshy bit that you dip into the cream and then suck/scrape it off with your teeth. Discard the leaf and repeat. As you get closer to the heart the leaves will become more tender and fleshy. Finally, when all the leaves have been worked over, you will be left with the heart and its hairy 'choke'. Using a spoon, scrape out and discard the hairy choke. You will be left with the delicious heart to eat last.

46

TUNA RED ONION SALAD WITH HARISSA YOGHURT DRESSING

I first wrote this recipe six years ago, when it contained a substantial quantity of croutons. I thought I was happy with it then, but over the years it's been tweaked, principally the carb-concentrated croutons have gone and consequently this is now a favourite lunch dish. Prepare everything in advance and then mix it all together at the last moment. Fiery North-African harissa, generally used to give couscous a kick, is widely available in cans and jars from many delis and better supermarkets and is a really very useful condiment to have in your store cupboard.

Serves 4 as a starter and 2 as a main course

4 ripe tomatoes
200g best-quality canned tuna in olive
    oil, drained and broken into chunks
200g fresh buffalo mozzarella, torn
    into small pieces
1 red onion, halved and thinly sliced
100g stoned black olives
1 lemon, peeled, segmented and
    chopped, removing the seeds
3 tbsp red wine vinegar
bunch of rocket, picked over
bunch of flat-leaved parsley, picked
    over

for the harissa yoghurt dressing
generous pinch of saffron
scant splash of boiling water
½ tsp harissa
2 canned anchovy fillets in olive oil,
    chopped
2 tbsp sherry vinegar or red wine
    vinegar
1 tbsp chopped fresh coriander
125ml olive oil

Blanch the tomatoes in boiling water for 20 seconds and then refresh them in a bowl of cold water. Remove the skins (which will now come away readily) and quarter the tomatoes. Place them in a bowl along with the tuna, mozzarella, onion, olives, lemon and vinegar. Mix well, cover and leave for 2 hours.

Make the dressing: put the saffron in a small bowl, pour on the boiling water and leave for 5 minutes. Add the remaining dressing ingredients except the oil and mix well with a fork. Then whisk in the oil. Adjust the seasoning, if necessary.

To serve: put the rocket leaves and the parsley in a salad bowl together with the marinated ingredients. Toss these together lightly and then pile up on 4 plates. Spoon the harissa yoghurt dressing over and around the salad.

ROAST FLAT MUSHROOMS WITH TALEGGIO AND PANCETTA

Childhood memories of picking large field mushrooms always spring to mind when I cook this dish. Sadly, I now have to rely on some Portobello mushrooms from the greengrocer in their place.

Serves 4 as a starter or 2 as a main course

4 large Portobello mushrooms
1 garlic clove, peeled
100g butter
8 slices of pancetta or smoked streaky
  bacon

200g Taleggio cheese
bunch of watercress, picked into leaves
1 large shallot, finely chopped
splash of Basic Vinaigrette (page 28)

Preheat the grill to its highest setting. Place the mushrooms, gill side up, on a grill tray. Cut the garlic into slivers and squash these pieces into the gills of the mushrooms. Melt the butter and drizzle it over the mushrooms, then season them with salt and pepper. Place the mushrooms under the grill and cook for 10 minutes or until they are cooked through. Take care not to burn them.

Meanwhile, heat a dry frying pan and fry the pancetta for a few minutes on each side until golden but not crisp. Set aside.

When the mushrooms are cooked, remove them from the grill and place slices of the cooked pancetta on top of each mushroom.

Cut the rind from the Taleggio and slice it into 4 pieces. If this proves tricky, then just cut it into small cubes. Arrange the cheese on top of the mushrooms. Place the mushrooms back under the grill, ensuring they are a few rungs down from the top. Grill until the cheese melts and starts to bubble. Your grill's efficiency will dictate how long this will take.

While the final cooking stage is underway, place the watercress leaves in a bowl together with the shallot. Season with salt, pepper, and add the vinaigrette. Toss well.

To serve, place the mushrooms on 4 plates and put a small handful of the salad on top of each. Serve immediately, as the heat of the mushroom wilts the watercress.

48

MEAT

You can fairly readily see that this is easily the biggest chapter in the book and, quite honestly, this is much more to do with my personal preferences than anything else. I am a lifelong committed carnivore … and I love meat in all its forms — be it beef, veal, lamb, pork, hams, poultry or game — and whether it is minced, as steak, in stews or roasts, made into sausages, or sometimes not even cooked.

Of course, I think it is beholden on me at this stage to make the point that when following a high-protein regime like this one, it is essential to watch your intake of saturated fat. So do make sure that as much as possible you favour lean meats, like pork fillet, skinless poultry and game, and also try to get a goodly proportion of your weekly protein from fish, which is low in saturates and full of healthy omega oils.

You will see that I have included a couple of recipes for veal dishes. Some people have issues with veal and, if you do, you can easily replace it with chicken or pork to great effect. I also love rabbit, which is a good food source, but not always easy to find these days and expensive when you can.

You may also note that there are not many recipes for chicken in this chapter (although there are several in Menus), as I think there is an abundance of recipes for it out there. I do, however, include several for duck — a particular favourite of mine. I know that it is fatty, but that's where much of the delicious taste lies. Like chicken, though, most of the fat content is in the skin, so you could always cook it with the skin on for the flavour and then discard the skin.

STEAK TARTARE

Good sourcing of ingredients is all with this dish. The food police will tell you that serving raw beef and raw egg on the same plate is more hazardous than a day trip to Chernobyl. However, buy the freshest organic eggs, go to a very good butcher and buy nice bright red fillet tails of beef (always cheaper than the rest of the fillet) and you minimize the risk. Yes, I am sure there are still associated risks with this dish (see my note on uncooked eggs on page 26), but I have been eating it for twenty years and have yet to suffer.

It's the classic garnish of pommes frites that scare me more. I only have to look at a bowl of the devils to feel the weight creep back on. A few of the thinnest slices of toasted sourdough or rye bread are a fabulous – but carb-laden – alternative. If you want to avoid carbs completely then scoop up the meat with sticks of celery or leaves of Little Gem lettuce.

Serves 4 as a starter

4 very fresh organic egg yolks
4 small cornichons
2 tsp capers
4 good-quality salted anchovy fillets
1 tsp Dijon mustard
1 tbsp tomato ketchup
splash of Worcestershire sauce

splash of Tabasco sauce
4 tbsp extra-virgin olive oil,
  plus more to serve
1 tbsp Cognac
400g beef fillet tails (see above)
2 shallots, finely chopped
1 tbsp chopped fresh parsley

Blitz all the ingredients except the beef, shallots and parsley to a coarse purée in a blender and set aside.

Cut the beef into long thin slices. Then slice these into long thin strips. Cut these strips into the smallest of dice. With your largest sharpest knife, give this pile of meat a good chop through to make it a little finer. Transfer it to a bowl and add the seasoning purée together with the shallot and parsley. Mix well and taste. Depending on personal preference, a little more Tabasco or Worcestershire sauce may be needed if you want it to have more of a kick.

To serve, shape the mix into 4 patties and place each in the middle of a plate. Drizzle a little oil around the plate and then sprinkle generously with black pepper.

STEAK WITH ANCHOVIES

Don't be scared of this one. The mention of anchovy often has the more timid eater running for cover, but their use here is subtle and complementary. Do this one on the barbecue and use rib-eye steaks – that extra bit of fat makes them all the more succulent.

Serves 4

4 good salted anchovy fillets, drained
grated zest and juice of 1 lemon
1 tbsp grainy mustard
1 garlic clove, crushed

200g unsalted butter, softened
4 rib-eye steaks, each about 250g
olive oil, for brushing
2 tbsp finely chopped shallot

Stoke up the barbecue and leave it to allow the coals to die down and glow intensely.

Chop the anchovy fillets finely and mix them in a bowl with the lemon zest, mustard and garlic, then mash this mixture into the butter. Season with plenty of pepper, taste it and adjust the seasoning. Set aside.

Season the steaks with salt and pepper and brush with olive oil. Grill them over the hot coals for about 3 minutes on each side. You want the exterior brown and caramelized and the interior juicy and rare.

Leave the steaks to rest in a warm place for 5 minutes and then serve on warmed plates. Spoon a dollop of butter on each steak and garnish them with a scattering of the chopped shallot.

PEPPERED STEAK WITH COGNAC

This is based on a steak I once had in Louisiana, which was served swimming in nut-brown butter and was one of the best I have ever eaten. It may all sound a bit too rich, but the substantial pepper crust cuts through this perfectly.

The times given below will produce a good rare steak. To cook them to medium, simply let the steaks rest in the oven without turning it off for 5 minutes and for 10 minutes if you want them well done.

Serves 2

2 fillet steaks, each about 230–250g
3 tsp cracked pepper
2 tbsp clarified butter

50–75g butter
3 tbsp Cognac

Preheat the oven to 100°C/gas ¼. Place the fillet steaks on a dish and press the pepper into one cut side of each steak only, pressing it into the meat with the heel of your hand to ensure it is well attached. Then season with salt.

Heat the clarified butter in an ovenproof frying pan and add the steaks, peppered side down, and cook briskly for 3–5 minutes, or until that first side is crusted and brown. Then turn over the meat and cook for 1 minute only.

Tip out the clarified butter from the pan and slip in half of the regular butter. Turn down the heat to a medium temperature and let the butter foam and cook to a gentle hazelnut colour. Baste the meat regularly with the browned butter. If the butter appears in danger of turning too dark then just lower the heat. Continue this process for 3–4 minutes.

Add the Cognac (if cooking over gas, then momentarily turn off the flame while doing this). It will splutter a little and, after thirty seconds, all the alcohol will have evaporated.

Remove the steaks from the pan and transfer them on a dish to the oven to rest for 8 minutes, turning off the oven as you put them in.

Transfer the steaks to warmed plates and add the juices that seeped out of the meat to the pan. Adjust the seasoning and then spoon the buttery juices over the steaks. Serve immediately.

FRIED COW SALAD AND SALSA

This dish is an indulgence and it is very much down to the home cook as to how the dish is finally presented. It is also important that you let go of European ideas of presentation and structure when applied to this dish.

My mother would be horrified by it. In fact she wouldn't even bother to cook it. After all, to cook meat so it falls apart definitely goes against the grain. However, when cooking for a generation of those who travel widely or have eaten in even the most ersatz of Mexican restaurants, then this dish shouldn't be too strange.

Vaca frita, or fried cow, is Cuban in origin. The original is a simple peasant dish served with beans or rice. The similarities stop there though. I have bastardized it and included some general South American flavours.

Serves 6

2kg chuck steak, trimmed and cut in two
olive oil, for frying

for the marinade
4 bay leaves
juice of 2 large lemons
3 tbsp red wine vinegar
4 medium onions
1 head of garlic, separated and peeled
1 bunch of coriander
3 tbsp sea salt
2 small red chillies, deseeded
6 allspice berries
2 tbsp cumin seeds, lightly toasted
100ml olive oil

for the salsa
6 tomatoes
2 small red onions, finely chopped
1 bunch of coriander, picked and
   chopped
2–4 chillies, deseeded (the quantity of
   chilli is down to their strength and
   your tolerance of chilli heat) and
   chopped
juice of 4 limes

to serve
250ml sour cream or crème fraîche
2 ripe avocados
1 iceberg lettuce

Day one: Trim any excess sinew from the surface of the meat. Place the meat in a bowl.
   Put the marinade ingredients in a blender and purée them, adding enough water to enable the machine to do its work. Pour the contents of the blender over the meat and combine it well. Cover and store in the fridge overnight.

Day two: Preheat the oven to 160°C/gas 3. Transfer meat and marinade to a suitably sized pot and add enough water so the meat is well covered. Place the pan on the stove and bring it to the boil. Skim away any of the foam-like scum that rises to the surface. Cover the pan with its lid and transfer it to the oven. Cook for at least 2–3 hours, or until the meat can be easily pulled apart with a fork.
   Remove the pan from the oven and let it stand to cool a little. With some caution, lift the meat from the pan on to a tray. Discard the cooking liquor. Using 2 forks, pull the meat apart into shreds. Place in a bowl, cover and refrigerate until needed.

Day three: Make the salsa, blanch the tomatoes in boiling water for 20 seconds and refresh in cold water. Skin, quarter and deseed the tomatoes. Cut them into medium-sized dice and place in a bowl with the onion and two-thirds of the coriander, reserving the rest for garnish. Add the chillies with the lime juice and a good seasoning of sea salt. Transfer to a serving dish.

Prepare the other accompaniments: put the sour cream in a serving dish. Halve the avocados, remove the stones and peel them. Dice the avocados and place them in a serving bowl. Separate the lettuce, separate and place the leaves in a serving bowl.

Heat some olive oil in a heavy non-stick pan and fry the shredded beef in batches until crisp yet still retaining some moistness. Place it in a dish at the table with the salsa and other accompaniments in separate bowls.

Consume as follows: take a lettuce leaf and spoon some beef into it, top it with salsa, sour cream, chopped coriander and a little avocado, roll it up and eat.

Note: in your never-ending quest for low-carb living you could, of course, embrace Cuba further and accompany the dish with a large Cuban rum – carb content zero.

SALT BEEF, SMOKED SAUSAGES, SAUERKRAUT & HORSERADISH

Cooking slowly and in a goodly quantity are the keys to this dish. Yes, it will make too much, but the leftovers are even better the next day. The sausages aren't the easiest to find unless you go on a day trip to France, when you can buy a good selection of smoked and unsmoked boiling sausages at most big supermarkets, take them home and freeze them for later use. Alternatively, search out some Polish sausages such as kielbasa. It does seem a ludicrous amount of sauerkraut, but it is very digestible.

Serves 8

2.5kg piece of salt beef (it must be brisket)
3 large onions, chopped
4 celery stalks, chopped
3 large carrots, chopped
2 leeks, chopped
a few black peppercorns
4 bay leaves
4 smoked boiling sausages, such as Montbéliard or kielbasa (see above)

1 large Morteau sausage
stick of fresh horseradish or jar of fresh grated horseradish (not creamed or sauce), to serve
for the sauerkraut
2kg cooked sauerkraut
8 juniper berries
300ml white wine
8 rashers of fatty bacon, sliced into little strips

Rinse the beef and soak it in a sinkful of cold water for 2 hours.

Place it in a large pan, cover it with cold water and bring to the boil. Tip out the water and add the onions, celery, carrots, leeks, peppercorns and bay leaves, then cover with water again. Bring to the boil and simmer gently. The beef will take 4–6 hours to cook, depending on its inherent toughness. You want it cooked until it is just beginning to fall apart. Top the pan up with water from time to time as necessary.

While the beef is simmering, prepare the sauerkraut. Place it in a pan with the juniper berries, white wine and sliced bacon. Mix well, cover and heat gently for about 45 minutes until it is piping hot.

When the beef is cooked, ladle some of the liquid into another pan and add both types of sausage. Simmer gently for 20 minutes to ensure these are cooked through. If they swell and appear to be on the verge of splitting, remove them from the heat.

Peel the horseradish and grate it into a bowl. It will make you weep profusely. Cover immediately with cling film to preserve its pungency.

This dish is best served in a 'Desperate Dan' sausage and mash style – i.e. piled high. Take a large deep serving platter and pile on the sauerkraut. Carefully lift the beef from its liquor and place it on top of the kraut. Arrange the sausages around the edge.

To serve, carve the meat and sausage on plates along with generous mounds of sauerkraut. Eat with the horseradish and perhaps some Dijon mustard.

BOILED SMOKED BACON WITH CELERY

This is a fine example of simple peasant cooking, given elegance by the wonderful flavour of celery. I couldn't believe my luck when I gave up carbohydrates and dishes like this were encouraged.

Serves 4

1.2kg piece of smoked streaky bacon, unsliced
2 onions, sliced
1 clove
¼ cinnamon stick
8 black peppercorns
2 heads of celery
100g butter

Rinse the bacon and place it in a pan of cold water. Bring it to the boil and then discard the water. Add the onions, clove, cinnamon and peppercorns. Cover with water again and bring up to a gentle simmer.

Chop the top third off the celery and add the trimmings to the bacon pot. Cook the bacon for about 2 hours, or until tender.

Preheat the oven to 180°C/gas 4. Split each trimmed head of celery lengthways. Melt the butter in a suitably sized roasting dish and, when it is foaming, add the celery, cut side down. Season lightly with salt and pepper, and cook steadily for a few minutes until the celery has coloured a little.

Draw off about 300ml of bacon cooking liquid and pour it over the celery. Cover tightly with foil, transfer the dish to the oven and cook for 35–45 minutes to ensure the celery is completely cooked.

To serve, place a half head of celery in each of 4 warmed shallow soup plates. Lift the bacon from its cooking liquid and carve slices from it to join the celery. Spoon some of the liquid over the bacon and celery.

59

VEAL 'CHEESEBURGER' IN A MUSHROOM BUN

Traditionally veal is often served with Livarot, not the easiest cheese to source. I love the richness Reblochon brings to the austerity of veal. Mind you, if you balk at veal, try using lamb and pairing it with Roquefort.

Serves 4

800g minced veal (90% lean, 10% fat)
finely grated zest of ½ lemon
1 garlic clove, crushed
2 tbsp chopped fresh parsley
dash of Tabasco sauce
2 egg yolks
1 small Reblochon cheese, preferably
  au lait cru

8 large bun-sized flat mushrooms
olive oil, for brushing
1 small red onion, thinly sliced into
  rounds
bunch of watercress, most of the stalks
  removed

In a bowl, combine the veal, lemon zest, garlic, parsley, Tabasco and egg yolks with a generous seasoning of sea salt. When well mixed, divide into 4 and shape into solid little patties. Transfer to the fridge and leave for 4 hours to allow the flavours to get friendly and the meat to set into shape. Remove from the fridge half an hour before they are needed.

Taking the Reblochon directly from the fridge, trim off the vertical circular edge, then cut the cheese into 8 wedges and store until required.

Preheat the oven to 180°C/gas 4. Pinch out the stalks from the mushrooms. Put the mushrooms in an ovenproof dish. Season and brush with olive oil. Transfer to the oven and bake for 10 minutes or until the mushrooms are just cooked. Set aside and keep warm. Preheat the grill to its highest setting.

Heat a cast-iron ridged griddle pan or, failing that, a large dry frying pan. Brush the 'burgers' with oil and cook for 5 minutes on each side over a medium heat. This will cook them right through should you be nervous of undercooked minced meat, but 3 minutes on each side will keep them nice and rosy pink inside.

Take 4 mushrooms and place them on a grill sheet and top each one with a 'burger'. Top each one in turn with 2 slices of Reblochon and place under the grill to melt the cheese. You only want the cheese to approach a fully molten state and certainly take great care to ensure it doesn't bubble up and colour.

Remove from the grill and place each portion on a serving plate. Scatter over the raw onion and some watercress, finishing off with the mushroom 'bun' tops. Serve.

VITELLO TONNATO

There is one traditional Italian dish that really stands out for me – not just for the flavour, but because it is ideal in that it won't take the cook away from the table when everyone else is eating. Vitello tonnato is a simple dish, cold roast veal with a sauce made from mayonnaise flavoured with tuna and some of the veal roasting juices. Ask your butcher to roll and tie up the veal for you.

You can make this dish even faster by using ready-cooked and sliced veal. Of course, you won't then have the cold meat juices, but a splash of water will work almost as well to thin down the sauce. If you don't eat veal, then try this sauce on some cold roast chicken, or if feeding a horde then you can even use it to dress cold turkey.

Serves 4 generously

1 piece of veal cut from the rump,
  about 1.25kg, rolled and tied
½ bottle of dry white wine
1 tbsp good-quality capers, to serve
  (optional)
1 tbsp chopped parsley, to garnish
  (optional)

for the sauce
200g canned tuna, drained
1 garlic clove, crushed
5 canned anchovy fillets in olive oil,
  drained, plus more to serve if you like
500ml mayonnaise
juice of ½ lemon
Tabasco sauce

Preheat the oven to 180°C/gas 4. Season the veal with salt and pepper and then brown it in some butter in a roasting pan over a moderate heat. Transfer to the preheated oven, pour in the wine and cook for an hour, basting frequently. Remove from the oven and let the meat go cold. Reserve the cooking juices for the sauce.

To make the sauce, put the tuna, garlic, anchovy fillets, mayonnaise and lemon juice in a food processor and blitz until smooth. Season with the Tabasco and salt, and then thin to the consistency of whipping cream with the gravy.

Thinly slice the cold veal and arrange it on serving plates, drizzle over the sauce and garnish, if you wish, with some good capers, anchovy fillets and chopped parsley.

SALTIMBOCCA ALLA ROMANA

A rather nice garnish for this classic Italian veal dish is to take four bottled baby artichokes, cut them in two and fry them swiftly in a little olive oil.

Serves 4

4 veal escalopes
8 sage leaves
4 slices of prosciutto di Parma

125g butter
125ml Marsala
lemon juice, to taste

Take two pieces of thick plastic cut from a clean heavy-duty carrier bag, wet them and place one piece on a solid work surface. On top of this, place a veal escalope and put 2 sage leaves on the meat. Cover this with a slice of prosciutto and then with the other piece of wet plastic. Proceed to beat the meat out so it doubles in size and becomes very thin. The water on the plastic stops it sticking too badly. Repeat the process until all 4 escalopes are beaten out.

   Divide the butter between two large frying pans and heat until it is foaming. Season the meat with pepper only and transfer them, ham side down, to the butter. Fry briskly for a minute or two and then turn the meat over. Cook for a further 30 seconds only and then add the Marsala to each pan. Cook for yet another 30 seconds only, then remove the pans from the heat.

   Lift the veal escalopes from the pans and place them on warmed plates. Tip the juices from one pan into the other, season with lemon juice and then spoon these juices over the veal. Serve immediately.

63

RABBIT WITH MUSTARD SAUCE AND BACON

Maille Dijon mustard is ideal for the sauce as it has a real kick to it. It is whisked in at the end just before the dish is served, so that the mustard retains all of its vibrancy.

Serves 4

knob of butter
a little olive oil (optional)
4 large rabbit legs
8 rashers of smoked streaky bacon
500g spinach, picked

for the mustard sauce
knob of butter
2 shallots, chopped
1 garlic clove, chopped
splash of white wine
100ml chicken stock
200ml whipping cream
2 tbsp Maille Dijon mustard (see above)

Preheat the oven to 180°C/gas 4. Make the mustard sauce: melt a knob of butter in a non-reactive pan and add the chopped shallots. Cook over a gentle heat until these are soft. Add the chopped garlic and cook for a further 30 seconds, then throw in the wine. Bring to the boil and reduce the volume of the wine by half. Add the stock and once again boil to reduce the volume of liquid by half. Pour in the whipping cream, bring to a gentle simmer and cook for about 5 minutes, or until the sauce has thickened ever so slightly. Season and set aside.

Heat a knob of butter in a heavy cast-iron dish until it is foaming. Season the rabbit legs with salt and freshly ground black pepper, and slip them into the butter. Cook over a medium heat and turn frequently to brown evenly. (If you are worried that your butter will burn, then add an equal quantity of olive oil to the pan when the butter is melting.) Transfer to the oven and cook for 20 minutes or until cooked. Remove them from the oven and keep warm.

While the rabbit is cooking in the oven, put the bacon rashers on a baking sheet and cook in the oven until crisp, then leave to rest in a warm place.

Remove the rabbit from its pan and allow to rest for 10 minutes in a warm place. Add the spinach and a seasoning of salt and pepper to the pan. Cook over a medium heat, turning frequently with a large spoon, until the leaves have wilted into the pan.

Press the excess liquid from the spinach and pile it on 4 warmed plates. Put a rabbit leg on each pile and garnish each with 2 rashers of crisp bacon. Finally – and with all haste – reheat the sauce and whisk in the Dijon mustard. Spoon this around the spinach and serve.

BREAST OF DUCK WITH OLIVE AND ALMOND SALAD

This dish is very Iberian in its feel. I use the common green olives stuffed with pimento for the salad, though it is worth spending a little extra for the best quality. They have a salty tang and lightness that is not too overpowering. In essence, the sauce is a simple aïoli seasoned with a pinch of cayenne and a splash of sherry. The quantities given here will make more than you need, but it is a most useful condiment to have to hand in the fridge.

Serves 4

4 duck breasts

for the salad
100g flaked almonds
2 tomatoes
200g pimento-stuffed green olives
2 shallots, thinly sliced into rings
small handful of picked flat parsley
   leaves
3 tbsp olive oil
juice of ½ lemon

for the sauce
1 large garlic clove
2 egg yolks
splash of sherry vinegar or red wine
   vinegar
300ml olive oil
2 tbsp Amontillado sherry

Preheat the oven to 160°C/gas 3. Make the salad: spread out the almonds on a baking sheet and bake in the oven for about 10 minutes or until golden. Remove from the oven and set aside to cool. Turn the oven setting up to 180°C/gas 4.

Blanch the tomatoes in a bowl of boiling water for 20 seconds. Refresh them in cold water and then remove the skins. Quarter them and remove the seeds, then cut them into small dice. Put in a bowl along with the almonds and all the other ingredients. Mix well and season with a good milling of black pepper. Set aside.

Make the sauce: crush the garlic to a smooth purée. Place it in a bowl along with the egg yolk and vinegar. Whisk together until the colour lightens. Then add the oil, drop by drop initially, building up to a confident thin stream. Finally whisk in the sherry and, if need be, a splash of warm water to give it the consistency of pourable cream.

Score the fat of the duck breasts with a sharp knife in a criss-cross fashion. Season well with salt and pepper. Place them, skin side down, in a heavy cast-iron ovenproof pan. Put it on the stove top and turn up the heat. When the duck has released some of its fat and the breasts are sizzling nicely, give them a quick basting and then put them in the oven for 8 minutes. Remove them from the oven, baste with the fat again and leave to rest in a warm place for a further 10 minutes.

To serve, place a pile of the salad on each plate. Slice each duck breast thinly and fan the meat out nicely against the salad. Finally drizzle over the sauce and spoon over any remaining meat juices that gather during the resting period.

FISH

I may love meat, but I also adore fish and seafood of all sorts. The sheer variety of delicate flavours and tender textures, combined with the speed at which they cook, are a gift to the chef.

There is much talk these days about superfoods and, if there are such things, then fish must be a prime candidate for the description. As well as being far and away the leanest source of protein and lowest in calories, fish are rich in a wide range of nutrients, notably the mineral selenium, which helps fight cancers and heart disease, as well as slowing the ageing processes. Oily fish, like salmon, mackerel, etc. are also full of the omega-3 oils which help fight everything from heart disease to depression.

Nutritionists have also just begun to find considerable scientific support for the age-old theory that fish is 'brain food'. Some of the omega-3 oils found in fish — and not in any other food — have been proved to have a demonstrable beneficial effect on learning ability.

In this country, people are rather strangely reluctant to explore fish other than the ones they know, but I really recommend that you try as many different types of fish as you can in recipes — if only for the fact that, because of natural resources being sorely depleted by overfishing, many of the varieties that we all know and love are soon to become virtually impossible to obtain. You don't have to stick to the type of fish I suggest in my recipes, you can try anything that has a fairly similar profile, i.e., white or oily, round or flat.

If you have previously only enjoyed your fish and shellfish breaded or swathed in floury batter, now is the time to discover its true flavour without any added carbs.

LEMON SOLE WITH BEURRE MAÎTRE D'HÔTEL

This flavoured butter makes an ideal freezer standby and is equally at home on a grilled rump steak as a portion of fish. The quantities given will make more than two portions, so cut the remainder into slices and freeze them individually so you can pull out the exact quantity you need a few minutes before you need it.

A good way of judging whether a piece of fish is properly cooked is to use the 'finger' method. Before you cook your piece of fish, prod it and remember the feeling. While cooking it, prod it gently a couple more times and you will begin to notice how the texture changes. When it is cooked it has a firm resilience that to me shouts readiness – when you prod it and your finger goes straight through you know you have overdone it.

Serves 2

2 whole lemon soles, each about 400–450g
melted butter, for brushing

for the maître d'hôtel butter
large bunch of parsley
250g unsalted butter, softened
grated zest of 1 unwaxed lemon
Tabasco sauce, to taste

To make the maître d'hôtel butter: pick the parsley and chop it as finely as you can. Place the softened butter in a bowl and beat in the chopped parsley together with the other ingredients and some sea salt to taste. Lay out a piece of cling film or greaseproof paper and place the butter in a sausage shape along the middle, then roll it up and twist the ends so as to give a neat tube shape. Store in the fridge for a couple of hours to set.

Preheat the oven to 180°C/gas 4. Brush a non-stick baking tray with a little butter and place the soles on it. Brush with a little more butter and season with salt and pepper. Cook for about 10–15 minutes or until they pass the test above.

Transfer to warmed plates, remove the butter from the fridge, unwrap and cut off 6 pencil-thick slices then arrange 3 on top of each of the fish. Serve straight away.

FILLETS OF PLAICE WITH GRILLED SPRING ONIONS AND CRAB VINAIGRETTE

Without experience, it is quite a palaver to pick the flesh from large whole crabs and buying dressed ones is a more cost-effective and time-saving way of obtaining fresh crab.

Serves 4

2 small dressed crabs
1 small red onion, finely chopped
1 small red chilli, deseeded and
  chopped
juice of 2 large lemons
1 tsp chopped fresh tarragon

1 tbsp chopped fresh parsley
olive oil, for brushing
100ml whipping cream
4 bunches of spring onions
4 portions of skinned plaice fillet, each
  about 180g

Lift the white meat from the crab shells and place it in a bowl together with the onion, chilli, juice of 1 lemon and the herbs. Season with salt and pepper and mix well. Add enough olive oil to loosen the mixture so it can be loosely poured.

Scoop out the brown meat from the shells and transfer it to a blender together with the juice of the remaining lemon. Blitz until smooth, add a splash of olive oil and the cream together with a good seasoning of salt and pepper. Give a final blitz to amalgamate and transfer to a bowl until needed.

Preheat a griddle pan and an overhead grill. Season the whole trimmed spring onions with salt and pepper and brush with olive oil. Cook them on the griddle pan until lightly charred and wilted. Keep warm.

Arrange the fish fillets on a non-stick baking sheet, season and brush them with oil. Grill for 5 minutes on one side only, or until cooked (the flesh just flakes).

Arrange a raft of spring onions on each of 4 warmed plates and then gently transfer the cooked plaice on top of these. Spoon over the white meat dressing and finally spoon the brown dressing around the edge.

HALIBUT WITH BRAISED LETTUCE

Sometimes when you entertain you need to introduce a little extravagance. I have been cooking variations of this dish for years and it is always very well received. It's a sort of coq au vin with fish. Prices of fish can fluctuate enormously particularly among the prime fishes like halibut and turbot. Some of the cheaper fishes will work in this dish with equal success – cod, pollack (a sort of cousin to cod), gurnard or very fresh grey mullet are excellent choices and dramatically cheaper. The lettuce softens and collapses, while its texture becomes remarkably delicate.

Serves 4

100g butter, plus more for brushing, for the sauce if necessary and to cook the fish
20 button onions, peeled
20 button mushrooms, stalks trimmed
100g streaky bacon, cut into small cubes

4 Little Gem lettuces
125ml red wine (Syrah/Shiraz give the best result)
400ml strong chicken stock
4 fillets of halibut, each about 180g, still with the skin
2 tbsp freshly chopped parsley

Preheat the oven to 180°C/gas 4. Melt the butter in a heavy pan over a gentle heat. Add the onions, mushrooms and bacon, season them lightly, cover and cook gently for 20 minutes until tender. Keep an eye on the pan to ensure the butter doesn't burn. Lift the cooked stuff from the pan and set aside.

Trim any loose leaves from the lettuces and cut each lettuce in half through the stalk. Place the halves, cut side down, in the pan and turn up the heat a little. Cook briskly for a minute, just enough to give them a little colour. Add the red wine and cook for a further 5 minutes, until the liquid is reduced by about half.

Return the onions, etc. to the pan together with the stock. Simmer gently for another 5 minutes to give a bit of body to the sauce. If it seems too thin, then add a good knob of butter and let it boil into the sauce. Keep hot while the fish is cooking

Butter an oven tray and lay the fish fillets on it. Season with salt and pepper and put a little piece of butter on top of each fillet. Transfer to the oven and cook for 5–10 minutes, depending on the thickness of the fish fillets, until when pierced with the tip of a knife the flesh inside has lost the opaque quality of uncooked fish flesh. Remove the fish from the oven and keep warm.

To serve, place 2 lettuce halves on each plate and put a fish fillet on top. Using a slotted spoon, lift out the garnish and let it fall haphazardly on each plate. Finally, adjust the seasoning of the sauce and spoon it over the fish. Scatter over the parsley.

PLAICE FILLETS WITH TOMATO AND CUMIN SAUCE

Here, instead of dry-roasting the spices, they are tempered in oil, which also brings out their aromatic qualities and transfers it directly to the oil. Leaving the skin on the fish fillets makes them easier to handle during cooking, but I do remove it myself as it make the fish much easier to deal with on the plate.

Serves 4

6 tbsp olive oil
¼ tsp cumin seeds
½ tsp black mustard seeds
8 whole black peppercorns
1 garlic clove, roughly chopped
1 small nugget of ginger, peeled and coarsely chopped

4 large very ripe tomatoes, roughly chopped
4 plaice fillets, each about 180g, still with the skin (see above)
fresh coriander leaves, to garnish (optional)

Put the olive oil, cumin, mustard seeds and black peppercorns in a medium non-reactive pan, heat them gently, jiggling the pan from time to time, and cook for about 5 minutes, taking great care not to overheat and burn the spices. Stir in the garlic, ginger and tomatoes, and season with sea salt. Turn down the heat and simmer gently for 5 minutes to ensure the tomatoes are cooked.

Transfer everything to a liquidizer and give it a good blitz. Push this sauce through a sieve to remove the spices and fibrous ginger. Return it to the pan, adjust the seasoning if required and keep warm.

Preheat the grill and oil the grill tray. Place the fish fillets on the tray, tucking their ends underneath themselves to give a nice plump cushion of fish. Brush them with oil and season them with salt and pepper. Put them under the grill and cook for about 5 minutes. If your grill isn't very powerful, it may take longer.

Place a pool of the tomato and cumin sauce in the middle of each of 4 warmed plates, carefully lift a fillet of fish from the grill tray and place it on top of the sauce. I don't think it needs anything else, but should you wish to embellish it any further then some fresh coriander leaves will work well.

HOT-AND-SOUR MONKFISH WITH CUCUMBER SALAD

I love spices, there is something about their intense and often medicinal flavour that draws me to them. None more so than tamarind, it has a unique sourness that when tasted neat repels but combined with other ingredients proves, for me, to be quite addictive. Tamarind is readily available in Oriental supermarkets and comes as a thick paste full of stones. It is soaked in hot water and then rubbed through a sieve. While this is the best stuff to use, you can cheat and buy a jar of tamarind purée that has had the stones and fibrous bits removed.

This recipe is based on a traditional curry from Goa. Be warned, the Goanese like their chillies, it is the home of 'Vindalho', that most delicious of dishes so sadly much bastardized in this country. This recipe, whilst pretty hot, is a revelation and is best served with some cooling accompaniments – minted yoghurt, cucumber salad or a vegetable dahl. This is one of the few dishes where the comforting and protective properties of rice may be needed.

Serves 4

1kg monkfish fillets (get your
  fishmonger to remove both layers of
  membrane)
4 dried red chillies
½ tsp cumin seeds
10 black peppercorns
1 tsp ground turmeric

8 large garlic cloves
golf-ball-sized piece of peeled fresh
  ginger
4 tbsp tamarind purée (see above)
2 large onions, sliced
3 tbsp vegetable oil
2 tbsp rice vinegar
2 tbsp chopped fresh coriander

Cut the fish into pieces, rub with some salt and set aside.

In a hot dry pan, toast the chillies, cumin seeds and peppercorns for 1 minute, then grind them to a paste with the turmeric, garlic, ginger and tamarind, using a pestle and mortar or small spice blender.

Fry the onions in the oil until golden, add the spice paste and cook gently for 5 minutes. Add the vinegar and cook until the liquid has evaporated. Add the fish and barely cover it with water. Bring to a gentle simmer, cover and cook until tender (pierce with the tip of a knife, the flesh inside should have lost the opaque quality of uncooked fish flesh).

With a slotted spoon, carefully transfer the fillets to a serving dish and keep warm. Simmer the liquor for another 10 minutes. If the liquid becomes too gloopy then add a splash of water. The fiery heat will remain but the spices will soften and mellow.

Finally, pour the sauce over the fish, scatter over the chopped coriander and serve.

FILLETS OF RED MULLET WITH WILD MUSHROOMS

Wild mushrooms aren't always that easy to get hold of outside of a good farmers' market. When you do manage to get some, spread them out on a tray and freeze them. When solid, transfer them to a freezer bag and store in the freezer until you need them.

Serves 4

good olive oil, for frying and brushing
200g wild mushrooms
2 shallots, sliced
½ garlic clove, chopped
1 tomato, skinned, deseeded and diced
3 tbsp good red wine vinegar
150ml dry white wine
3 tbsp chopped parsley
4 large fillets of red mullet

Heat a generous amount of olive oil in a frying pan and fry the mushrooms for about 4–5 minutes, or until the mushrooms have coloured lightly. Season them well with salt and pepper, then add the shallots followed by the garlic and tomato. Pour in the vinegar and cook until it has all but evaporated. Add the wine, bring to the boil and simmer gently for 5 minutes. Remove from the heat, adjust the seasoning and stir in the parsley.

While the mushrooms are cooking, preheat the grill. Place the fish fillets on the grill tray, skin side uppermost, brush them with some olive oil, sprinkle with a little sea salt and grill for about 5 minutes without turning. The skin will crisp and blister beautifully.

To serve, arrange the red mullet fillets on 4 warmed plates and spoon the mushrooms over them or alongside.

POTTED SHRIMPS

Make these the day before you need them, as I have found that without a little rest the spicing can prove to be a little harsh. The power of the combination of cayenne and black pepper with Tabasco does make a great difference. The shrimps can be ordered from a good fishmonger.

Serves 4

250g unsalted butter
½ tsp freshly ground mace
¼ tsp cayenne pepper
500g peeled brown shrimps

Tabasco sauce
leaves from Little Gem hearts, to serve
lemon wedges, to serve

In a wide shallow pan, melt the butter and then add to this the mace and cayenne pepper. Cook the spices gently for 2 minutes, but take great care not to allow the butter to colour.

Add the shrimps and cook for a further 5 minutes or so. The shrimps are already cooked, so it is important that you just heat them through properly. Finally, add a seasoning of Tabasco sauce and also a little sea salt and some black pepper.

With a slotted spoon, lift out the shrimps and pack them into 4 little pots or ramekins. Reheat the remaining butter, skim off any bits and then pour the butter over each pot of shrimps to seal them. Transfer to the fridge, but remember to take them out of the fridge 40 minutes before you want to serve them.

Serve with Little Gem heart leaves and lemon wedges.

HERRING AND SMOKED BACON WITH FENNEL, FRENCH BEAN AND TOMATO SALAD

Alsace bacon, which has a dark smoky flavour reminiscent of kippers, can always be found on trips to the hypermarket in France but if this isn't practical then use a strong traditional English type.

Serves 4

250g French beans, topped and tailed
4 tomatoes
1 fennel bulb
2 shallots, peeled
olive oil, for frying
8 rashers of smoked bacon, preferably
  Alsace (see above)
8 fresh herrings, filleted but still with
  skin
4 tsp chopped fresh parsley

for the dressing
2 egg yolks
1 garlic clove, crushed
6 large canned anchovy fillets in olive oil
½ tsp English mustard powder
1 tsp Maille Dijon mustard
splash of Tabasco sauce
splash of Worcestershire sauce
250ml olive oil
juice of 1 lemon

Make the dressing: in a liquidizer, blitz the egg yolks, garlic, anchovy and both mustards with the Tabasco and Worcestershire sauces. When well combined, pour in the olive oil in a slow steady stream, as for mayonnaise. When the emulsion becomes too thick (the consistency should be of barely pourable mayonnaise), add the lemon juice. Adjust the seasonings and, if still overly thick, add a little hot water.

Salt the boiling water and boil the beans in it until properly cooked. They must retain a vague resilience but should definitely not be crunchy. Remove them from the water and plunge them into a sinkful of very cold water to stop them cooking.

Blanch the tomatoes for 10 seconds or so to loosen the skins, refresh them in cold water and then skin and quarter them. Remove the seeds and cut the tomatoes into strips. Trim the tops of the fennel bulb and cut it in half. Cut off the bottom centimetre and then slice the bulb from top to toe into fine strips. Halve the shallots as well and cut them into similar strips. Mix all of the vegetables in a bowl with the dressing and set aside for 30 minutes.

Heat some olive oil in a large frying pan, place in the bacon rashers and cook gently until lightly coloured but not crisp. Remove and keep warm.

Season the herring fillets and fry them, skin side first, in the hot bacon-flavoured pan until they are sizzling away, about 2–3 minutes. Turn them over and leave for a few more seconds. This will give you a barely cooked herring fillet which is ideal if they are spankingly fresh.

To assemble, divide the salad into 4 tight piles, one on each plate. Then stack up 2 herring fillets and 2 rashers of bacon on top and garnish with the chopped parsley.

BAKED CRAB WITH TOMATO AND TARRAGON HOLLANDAISE

This is really very, very rich... very, very decadent... Do I really care – no, not when it tastes quite this good.

Serves 4

125g spinach leaves, stalks removed
¼ garlic clove, crushed
1 tbsp crème fraîche
knob of butter
300g white crab meat (or 4 dressed crabs if you want to bake crab in the shell; use brown meat in sauce for fish)
Tabasco sauce
3 tbsp finely chopped chives

for the tomato and tarragon hollandaise
2 tomatoes
200g unsalted butter
knob of butter
2 egg yolks
juice of ½ lemon
small packet of fresh tarragon leaves, picked

Rinse the spinach several times to remove any traces of grit. Bring a large pan of salted water to the boil and blanch the spinach for 30 seconds. Drain in a colander and return to the pan together with the garlic and crème fraîche. Cook for 2 minutes, then put it in a food processor and blitz it to a purée. Transfer to a small pan and set aside.

Prepare the tomatoes for the sauce: blanch them in boiling water for 20 seconds, refresh them in cold water and then remove the skins. Quarter the tomatoes, remove the seeds and then cut the tomatoes into small dice. Set aside on a piece of kitchen towel to absorb any excess moisture.

Make the tomato and tarragon hollandaise: melt the 200g butter in a saucepan until hot. Place the egg yolk and lemon juice in a metal bowl that you have set over a pan of gently simmering water (not touching it). Whisk until the mixture thickens and goes pale. Turn off the heat under the water. If at any point the sauce starts to go grainy and curdle, then add a tiny splash of boiling water from the kettle. Whisk in the hot butter as you would for mayonnaise, that is to say in a very slow steady stream. Keep the kettle to hand should the sauce start to split. When all the butter has been incorporated, remove the bowl from above the hot water and season with salt and pepper. Chop the tarragon and stir it in together with the chopped tomato.

Preheat the grill. Melt the butter in a pan and add the crab, stir for 3–5 minutes, or until piping hot. Season with Tabasco and sea salt, and add one-third of the chives.

Divide the spinach purée between 4 small shallow heatproof dishes or crab shells and then pile in the crab. Spoon the tarragon hollandaise over the top. If it is too thick to ladle over the pile of crab, then reach for the kettle again and whisk in a little more hot water. Put under the grill, for just long enough to colour the hollandaise slightly, this will take about 30–60 seconds.

With a suitably thick cloth or oven glove, transfer each dish or shell to a plate that you have lined with a napkin, scatter over the remaining chives and serve.

AVOCADO, CRAB AND CUCUMBER SALAD

In this simple and flavourful salad, the soy sauce brings out the taste of the avocado in a way that salt fails to do.

Serves 4

½ cucumber
2 tsp salt
1 large ripe avocado
1 shallot, roughly chopped
juice of 1 medium lemon

Tabasco sauce to taste
about 3–4 tbsp Kikkoman soy sauce
2 spring onions
400g white crab meat
fruity extra-virgin olive oil, for drizzling

Peel the cucumber and split it lengthways. Using a teaspoon, scrape out the seeds and discard them. Slice the cucumber at a slight angle and at the thickness of 2 matchsticks or thereabouts. Place the slices in a colander, mix in the salt and leave them to degorge their slightly bitter juices.

After 30 minutes, rinse the cucumber slices and soak them in cold water, which you need to change regularly, for an hour. When they no longer taste salty, they are ready and can be drained and set aside.

Halve the avocado, discard its stone and scoop the flesh into a blender. Add the shallot, half the lemon juice, a splash of Tabasco and 3–4 tablespoons of soy sauce. Blitz until smooth, adding just enough cold water to enable everything to purée properly. Adjust the seasoning, adding a little more soy sauce if needed. Set aside.

Trim the spring onions, cut off the green tops and reserve them. Slice the white parts at a sharp angle and place them in a bowl with the cucumber. Drizzle in a little olive oil and the remaining lemon juice, season lightly and mix well.

Spoon the avocado purée into the middle of 4 plates. Arrange a bed of the cucumber/spring onion salad on top of this. Divide the crab into 4 equal piles and put one atop each of the beds of salad. Drizzle some olive oil around the plate (between the raised edge of the plate and the salad itself) and then drip some drops of soy sauce to sit in the oil. (When you spend this kind of money on the crab it really is worth embellishing it properly.) Finally, slice the reserved green parts of the spring onions as finely as possible, in much the same way as you would chives, and then scatter them over the whole show.

PRAWNS WITH SHERRY, CHILLIES AND GARLIC

This is a close variation on the Andalusian classic gambas pil pil. *Leaving the shell on the prawns gives them a greater depth of flavour and I love getting stuck in to them with my fingers and the mess that is associated with that. It may be a clash of cultures, but a good bowl of Greek salad to accompany the prawns is a winner. Buy a half bottle of manzanilla or fino sherry and chill it well. What you don't put into the prawns should be put into glasses and drunk along with the prawns.*

Serves 4 as a starter or 2 as a main course

3 tbsp extra-virgin olive oil
500g raw prawns, headless but still in the shell
2 garlic cloves, thinly sliced

1 small red chilli, deseeded and cut into strips
1 tbsp chopped flat-leaf parsley
1 tbsp chopped coriander leaves
3 tbsp manzanilla or fino sherry
lemon wedges, to serve

Heat the olive oil in a large cast-iron skillet or frying pan. When hot, season the prawns and slide them into the pan. Spread them out into a single layer and fry fiercely for 2 minutes, then turn them over and cook for a further minute.

   Scatter over the garlic, chilli and herbs, shake the pan and add the sherry. When the spluttering subsides, serve immediately with wedges of lemon.

GRILLED SQUID, CLAM, MUSSEL AND RADISH SALAD

This makes a great 'get your fingers messy' dinner party starter. It is important to buy the freshest of squid and shellfish for this dish as it is so lightly cooked.

Serves 4 as a generous starter or 2 as a main course in summer

1kg clams, Venus or palourdes for preference
1kg mussels
3 tbsp olive oil
2 tbsp chopped shallots
2 garlic cloves, finely chopped
1 red chilli, deseeded and chopped
250ml dry white wine
1 tbsp chopped parsley
1 tbsp chopped coriander
1 bunch of radishes
500g cleaned squid
1 bunch of chives, thinly sliced
1 large lemon, to serve

for the aïoli
2 egg yolks
pinch of cayenne pepper
½ tsp Dijon mustard
200ml fruity olive oil
splash of sherry vinegar
splash of Amontillado sherry

Rinse the shellfish in cold water. Heat 2 tablespoons of the oil in a large pan, add the shallot, garlic and chilli, and cook gently until softened. Add the mussels and clams together with the white wine. Clamp on the lid tightly and turn up the heat. Cook for several minutes until the majority of the bivalves have opened. It is preferable to let a few remain unopened to avoid those that open more easily overcooking.

With a large slotted utensil, lift the shells out into a bowl. Bring the cooking liquor to the boil and reduce by half. Transfer to a clean bowl and let cool. Fold in the herbs.

Top and tail the radishes, slice them thinly and fold into the cooking liquor bowl. If the mixture seems a little dull add another splash of olive oil. Season with salt and pepper if required. Pour this dressing over the mussels and clams, and mix together.

To make the aïoli, combine the egg yolk, cayenne pepper, mustard, vinegar and a seasoning of salt to taste. Whisk well and then add the oil in a thin steady stream. When it gets too thick and appears on the verge of curdling, beat in a little sherry. Finally add another small splash of sherry, but not enough to make the aïoli lose its body.

Cut the squid into small squares and score it vertically and horizontally on the exterior surface with the sharpest of knives (ask your fishmonger to put one scoring line on the outside surface so you can tell which it is when you get home).

Lightly oil a griddle pan, wok or similar and preheat it. Season the squid and cook it in small quantities so it just curls up. Lift out and place in a bowl. Mix in the chives.

To serve, place a pile of the shellfish on each plate and garnish with the squid pieces, finishing it off with a dollop of aïoli and a lemon wedge.

EGGS

Eggs are just the most brilliantly compact, prepacked food source. Moreover, they are cheap, satisfying, tasty, high in protein and very nutritious. Just the thing for a high-protein, low-carb regime. Eggs are also a gift to the cook in that they are extraordinarily flexible and can be used to make things as diverse as scrambled eggs, omelettes, batters, cakes and unctuous sauces like mayonnaise and hollandaise.

They also offer an unparalleled vehicle for other flavours, working well with a wide range of herbs, especially chives and chervil. Omelettes can be wrapped around an almost limitless array of fillings, containing meat, fish and vegetables and cheese.

They are undoubtedly among the most consoling of comfort foods. A perfectly boiled or poached egg is one of life's great treats. Swathe it in a sauce or add it to a salad, fish or vegetable dish, so that the softish yolk breaks over the other ingredients and you have a true delicacy.

I do also say it here and there in the book where appropriate, but I think this is definitely the place to remind you again of the inadvisability of serving raw or very lightly cooked eggs to children, the elderly, pregnant women and invalids, because of the dangers of salmonella poisoning.

I also recommend that as much as possible you spend those few extra pennies to buy the best organic, free-range eggs that you can get your hands on – you will really notice the superior flavour. Also, as you will be using your eggs so rapidly on this regime, don't bother to store them in the fridge but instead simply keep them in a cool place. They do perform much better in several preparations when they are at room temperature.

OEUFS EN MEURETTE

This dish is, in essence, boeuf bourguignonne using eggs in place of beef. Had you any gravy left over from a stew enriched with red wine, then this would make an ideal basis for this sauce. In place of such good fortune, this simple version works very well indeed, providing you make your own dark fresh chicken stock and reduce down until it is strong and flavourful. Traditionally, these eggs are served on bread croûtes, but too bad...

Serves 2

2 knobs of butter
2 tbsp chopped shallot
250ml heavy red wine
250ml strong chicken stock
50g lardons

10 button onions, peeled
10 button mushrooms
4 of the freshest eggs
splash of white or red wine vinegar
1 tbsp chopped parsley

Melt one of the knobs of butter in a non-reactive saucepan. Add the shallot and cook over a medium heat until soft. Add the red wine, bring to the boil and reduce by half. Add the chicken stock and reduce again by half.

In another pan, melt the remaining butter and add the lardons, onions and mushrooms. Season them and initially cook quite briskly to give them an attractive brown colouring. Then add the red wine sauce and simmer gently until the vegetables are properly cooked through. Season and keep hot.

To poach the eggs, fill a large pan with lightly salted water, add a splash of vinegar and bring to the boil. Break each egg into a small vessel and then slip them in one by one. Cook over the barest of simmers until the whites have set firmly.

Carefully lift out the eggs and place two of them in each of two soup plates. Spoon over the sauce and garnish with a generous sprinkling of chopped parsley.

DUCK CONFIT AND GOATS' CHEESE OMELETTE

Originally a vegetarian dish using girolles in place of the duck, it was general lack of availability of wild mushrooms that brought me to introduce a meat element to this simple recipe. Yes, it is a little richer but it does make the perfect filling supper dish.

Serves 2

1 confit duck leg
½ garlic clove, finely chopped
1 tbsp chopped parsley
6 free-range eggs

2 fresh young goats' cheese, such as
  crottins
bunch of fresh chervil, picked

Remove the confit duck leg from its covering fat and leave it to stand at room temperature for 30 minutes.

Preheat the oven to 180°C/gas 4. Place the duck leg skin side down in a small heavy ovenproof frying pan. Using the palm of your hand, squash the leg down into the pan to bring as much of the duck's skin into contact with the pan's surface as possible. Heat until the duck starts to sizzle lightly and then transfer the pan to the oven.

Cook the duck in the oven for 20 minutes, or until the skin has crisped to a good crunchy golden-brown. Remove the pan from the oven, lift the duck leg on to a plate and allow to cool to lukewarm. Strain the fat into a bowl and, when it is cold, store in the fridge for later use.

Carefully lift the duck skin from the meat and cut it into strips. Prise the meat from the bones and shred it into a bowl. Mix in the garlic and parsley.

Break 3 eggs into each of 2 bowls. Season each with salt and pepper and beat them. Crumble a goats' cheese into each bowl and gently stir into each half of the duck meat and skin together with half of the chervil leaves.

Heat a scant quantity of duck fat in a non-stick or well-proved frying pan. Pour in the contents of one of the bowls and, using a fork, stir the egg mixture as you would for scrambled eggs. When the mixture is looking halfway cooked and vaguely creamy, stop stirring and allow the omelette to set.

Using a spatula, tip the pan and fold over the omelette to form a nice plump cushion, then tip it on to a plate. With all swiftness, repeat the process with the remaining omelette mixture. Serve with a large salad.

SOFT BOILED EGG, SMOKED COD'S ROE AND AVOCADO SALAD WITH HORSERADISH DRESSING

This salad is pretty aggressive in flavour and seems like a collage of very rich ingredients. Don't worry, though, as the bite of horseradish and cracked pepper makes the perfect foil in a well-acidulated vinaigrette.

Serves 4

4 free-range eggs, brought to room temperature
1 frisée lettuce
1 ripe avocado
50g macadamia nuts
300g smoked cod's roe
small bunch of chives

for the horseradish dressing
2 shallots
100ml olive oil
juice of 1 lemon and grated zest of ½
2 tsp grated horseradish, preferably fresh
1 tsp cracked peppercorns
a little Maldon sea salt

Bring a pan of salted water to the boil and lower in the 4 eggs. Bring the pan back to the boil and cook the eggs for 6 minutes exactly. Lift them out and plunge them into a bowl of iced water, then set aside for half an hour to allow them to cool right down.

Separate the frisée leaves and put them in a large salad bowl.

Make the horseradish dressing: put the chopped shallot into a bowl together with the olive oil, lemon juice and grated zest, horseradish, cracked peppercorns and sea salt. Stir everything together and set aside.

Halve the avocado, peel it, remove the stone and dice the flesh, then add the diced flesh to the salad bowl.

Put the nuts in a plastic bag and give them a gentle bashing to break them into coarse nibs. Add them to the salad bowl.

Slice the cod's roe and shell the eggs. Dress the salad with four-fifths of the dressing and toss well. Divide it between 4 plates and place 2 egg halves on the top of each pile, then arrange the cod's roe slices around the egg. Spoon the rest of the dressing over the eggs. Chop some chives and scatter them over the top.

EGGS FLORENTINE

This is the type of dish that would have the avid diet junkie running for cover. Classic and delicious as it is, the weight still disappears.

Serves 2

125g spinach leaves, stalks removed
¼ garlic clove, crushed
1 tbsp crème fraîche
4 tsp red wine vinegar
4 of the freshest large eggs
3 tbsp finely chopped chives

for the hollandaise sauce
150g unsalted butter
2 egg yolks
juice of ½ lemon

Rinse the spinach several times to remove any traces of mud. Bring a large pan of salted water to the boil and blanch the spinach, then drain in a colander. Squeeze out the excess liquid and place the spinach on a chopping board and chop it to a coarse pulp. Return it to the pan together with the garlic and crème fraîche and cook it gently for 2 minutes. Adjust the seasoning and set aside.

To make the hollandaise sauce: melt the butter in a saucepan until hot. Place the egg yolks and lemon juice in a metal bowl set over a pan of gently simmering water (not touching it). Whisk until the mixture thickens and goes pale in colour. If it starts to go grainy and curdle, then add a tiny splash of boiling water from the kettle. Whisk in the hot butter as you would for making mayonnaise, that is to say in a very slow steady stream. When all the butter has been incorporated, remove the bowl from the heat and season the sauce with salt and pepper.

Preheat the grill. Heat a panful of water and add the vinegar and a little salt. Bring to the gentlest of simmers. Break the eggs into small cups and slip each egg into the simmering water. Cook the eggs for about 4 minutes – ideally the whites should be just set and the yolk still runny. Remove the eggs from the water using a slotted spoon and set them to rest on a piece of kitchen towel.

Heat 2 small ovenproof dishes under the grill for a minute. Place a layer of the spinach on the bottom of each dish followed by a pair of eggs. To finish, spoon the hollandaise sauce over the eggs and stick the dishes back under the grill for 10–30 seconds, or until the sauce has glazed to a delicate pale brown. Sprinkle over the chopped chives to serve.

EGG 'PANCAKES'

I first came up with these little beauties cut into strips as a garnish for a crab salad. In fact, they make a very nice addition to quite a lot of mixed salads. Where they really excel, though, is as unusual wraps for delicate little fillings, such as smoked salmon and cream cheese. In such a case, dill would be first-class as the added herb.

Makes 12

olive oil, for frying
4 large free-range eggs
1 tbsp crème fraîche or sour cream

2 tbsp chopped mixed fresh herbs
(I like to use parsley, chives, chervil and
    tarragon but really it's up to you)

Preheat the grill to high. Brush a non-stick pan which has an ovenproof handle with a little of the olive oil.

Break the eggs into a bowl and beat them well with the crème fraîche and a good seasoning of salt and pepper. Stir in the herbs.

Place the pan over a highish heat and, when it is hot, add enough egg to cover the base of the pan, a covering barely thicker than your average pancake. It will colour on the underside and set in as little as 30 seconds.

Then hold the pan under the grill for 30 seconds to set the top firmly. Lift it out on to a piece of non-stick greaseproof paper and repeat the whole process until all the egg mixture has been cooked, wiping out the pan with kitchen paper and adding a little more oil as necessary. Use the 'pancakes' as you see fit (see above).

PICKLED EGG AND CRISP BACON SALAD

Pickled eggs are usually found swimming in a dark brown pickle on the counter of a fish and chip shop, and are best avoided. Garners make an excellent example that are wonderful in this old-fashioned vegetable salad. The cream dressing is in place of sugar-laden salad cream.

Serves 2

8 rashers of the thinnest-cut smoked
    streaky bacon
4 celery stalks
2 large spring onions
12 cherry tomatoes
4 button mushrooms
splash of olive oil

100ml crème fraîche
2 tbsp red wine vinegar
1 tbsp Worcestershire sauce
1 tsp Dijon mustard
3 pickled eggs, Garners for preference
    (see above)

Preheat the grill. Lay the bacon rashers out on the grill pan and cook under the grill, turning once, until golden and crisp. Lift on to a piece of kitchen paper and set aside.

Heat a large pan of water. With a vegetable peeler, remove the stringy part of the celery and then slice it at an angle to create nice crescent-shaped pieces. Slice the spring onion whites at a similar angle, but reserve the top green parts.

Blanch the cherry tomatoes in the boiling water for 15 seconds and then refresh them in cold water. Carefully skin them. Slice the mushrooms thinly.

Put all these vegetables into a bowl, then gently mix them together, along with the splash of olive oil and a seasoning of sea salt and pepper.

Combine the crème fraîche, vinegar, Worcestershire sauce and mustard, and season the mixture with salt and pepper. If it thickens too much, then thin it down with a splash of hot water.

Pile the salad on 2 plates. Cut the eggs in half, put 3 halves on each plate and then drizzle over the cream dressing. Thinly slice the reserved spring onion greens and scatter them over the whole affair.

CHEESE

I have never been able to give up cheese when previously attempting to assert some form of control over my ample form – it is just too addictive. Real farmhouse-made cheeses are among the few foods that are still alive as we eat them. They are a carefully nurtured and matured food source that invariably improves with age and really does exhibit the tradition and love of the cheese-maker for his raw ingredients. All this as well as the simple fact that cheese in its various incarnations is quite simply delicious. 'Fat content' are the two words that are invariably uttered by many of you who question the diet's efficacy. Consuming fat - whether it be dairy-, vegetable- or animal-based is, however, one of the best tools around for losing weight. It stops you feeling hungry, therefore those cravings for carbohydrate disappear, and if you consume fat in place of carbohydrate then it will actually speed up the burning of stored fat. It really does sound too good to be true, but the proof is here. I ate nothing but three of the cheese recipes that follow for a twenty-four-hour period when I was testing them. Admittedly it was quite a bit more cheese than I would normally consume in that period, but when I got on the scales the next day I was already a pound lighter than the day before.

As with all things in life, show a little common sense. If you make a cheese a substantial part of a given meal then don't eat it for the next few meals. Issues such as blood cholesterol levels and heart disease will come to the fore – and quite rightly so. Just skimming through the book and only taking in the parts that you want to, would be irresponsible. You have to know what you are doing to your body to avoid any possible detrimental effects to your health.

BAKED COOLENEY WRAPPED IN BACON

The wrapping of the cheese in foil is all important, the smallest hole and your resultant cheese sauce will leak out and disappear in a cloud of smoke. This dish came into being as a result of looking at some Irish classic ingredients and then marrying this delicious Irish cheese with the Alpine dish of boite chaude, *in which a Vacherin Mont d'Or is baked in its box until liquid. If you can't find any Cooleney, then a small Reblochon or Camembert will do just as well.*

Serves 2

50g butter, plus more for smearing
   the foil
½ Savoy cabbage, finely shredded
1 small garlic clove, peeled

1 mini Cooleney cheese
1 tsp Maille Dijon mustard
6 slices of green streaky bacon
3 tbsp white wine

Preheat the oven to 180°C/gas 4. Melt the butter in a medium pan and add the cabbage. Season with salt and pepper, place the lid on the pan and cook over a gentle heat, until the cabbage is softened and the excess liquid has evaporated.

Quarter the garlic lengthways and stud the cheese with the 4 pieces. Spread the mustard over the top of the cheese and then wrap the cheese in the bacon. Take a square of kitchen foil and smear it with butter. Place the cheese in the middle, season with pepper and scrunch up the foil around it, leaving a hole at the top so you can then spoon in the wine. When the wine is added, pinch the top together, put the cheese in the oven and bake for 20 minutes.

To serve, put the cooked cabbage in a warmed serving dish and then carefully open up the foil parcel. Lift out the bacon-wrapped cheese, then place it in the middle of the cabbage. There will be a lot of runny cheese at the bottom of the foil, carefully pour this over the baked cheese and cabbage. Finish by milling over some black pepper and serve while piping hot.

WELSH RAREBIT OMELETTE

There are certain favourite dishes that I felt that I couldn't walk away from. A savoury, almost soufflé-like topping of cheese on a slice of fresh toast is a Sunday-night supper dish that I still struggle at times to hold at bay. Luckily, my love of omelettes was the saving grace. To take the key ingredients and fold them into some freshly beaten eggs became a new dish and one that happily stands comfortably on its own. A genuine farmhouse Cheddar, with that sort of assertiveness that seems to tingle on the roof of your mouth, is vital here – Quicke's, Montgomery or Keen's are perhaps the finest examples.

Serves 1

3 free-range eggs
½ tsp Dijon mustard
¼ tsp English mustard
1 tbsp Worcestershire sauce

dash of Tabasco sauce
100g best mature Cheddar (see above), grated
15g unsalted butter

Break the eggs into a bowl and beat well. Stir in the remaining ingredients, except the butter, season with salt and a good milling of black pepper, and mix together.

Heat the butter in a non-stick or well-proved pan and, when foaming, pour in the egg mixture. Using a spatula, move the mixture about so it forms curds of set egg. When most of the egg has set into curds, leave the pan well alone for a further minute to let the mixture set. Then tilt the pan and fold the omelette over.

Slip it out on to a plate and serve immediately.

MELANZANE ALLA PARMIGIANA

For this good 'meaty' vegetable dish I use canned tomatoes to get the most flavourful result. I was given this recipe on a trip to Apulia. A layer of mortadella is usually included there as it seems that, while many Italians regularly don't eat meat in a meal, the concept of vegetarianism and choosing to exclude meat completely from your diet is a strange one. The mortadella, being lightly smoked, is a great enhancement to the dish, but I have left it out here to keep the recipe meat-free.

Serves 4

3 large or 4 medium aubergines
flour, for coating
4 eggs
splash of milk
olive oil, for frying
2 balls of good-quality mozzarella, torn
   into shreds
bunch of basil leaves, picked and torn
75g grated Parmesan cheese

for the tomato sauce
3 tbsp olive oil
1 onion, finely chopped
1 bay leaf
1 tsp fresh oregano, chopped
4 garlic cloves, chopped
4 large (400g) cans of chopped plum
   tomatoes

Preheat the oven to 180°C/gas 4. Make the tomato sauce: heat the olive oil in a large non-reactive saucepan. Add the onion, season with salt and pepper and cook over a medium heat until soft. Add the bay leaf, oregano, garlic and canned tomatoes, bring to a gentle simmer and cook for 1 hour. Stir regularly and, when the sauce starts to turn too thick and claggy, let it down with an appropriate splash of water. It should be pretty sloppy in consistency when finished. Adjust the seasoning if necessary.

While the sauce is simmering, slice the aubergine to a thickness of 1 centimetre. Put the flour in a shallow dish and season it well. Break the eggs into a dish, add a good splash of milk, season and beat. Heat some olive oil in a wide pan. In batches, dip the aubergine slices into the flour, followed by the beaten egg, and then fry them until golden on both sides. Remove them from the oil and drain on some kitchen paper.

Spread a little of the tomato sauce over the bottom of a suitable baking dish. Then put in a layer of aubergine, followed by more sauce and a scattering of mozzarella and basil leaves. Repeat this process until the dish is full. Finally sprinkle over the grated Parmesan.

Transfer the dish to the oven and bake for 45–60 minutes, or until the contents of the dish are bubbling away and piping hot. Serve.

GRILLED MARINATED HALLOUMI

Ideal for vegetarians, this dish uses Greek halloumi, the dense meaty cheese with a fabulous salty tang. Cook it on a ridged cast-iron grill or, in summer, as part of a barbecue. Surprisingly, it doesn't melt. It is best to start the marination the day before. This also makes a great accompaniment for the grilled lamb cutlets and tahini dressing on page 42.

Serves 2, 3 or 4 as a starter

200g(ish) packet of halloumi
1 small garlic clove, chopped
generous pinch of dried oregano

grated zest of ¼ lemon
generous pinch of chilli flakes
extra-virgin olive oil

Cut the cheese into 8 slices – they should each be just under a centimetre thick – and lay them out in a dish, preferably in a single layer. Sprinkle the garlic, oregano, lemon zest and chilli flakes over the cheese, then drizzle over just enough olive oil to enable you to ensure all the surfaces of the cheese are covered. Cover with cling film and store in the fridge overnight.

Next day, heat a griddle pan. Lift the cheese slices from the marinade, letting any excess oil drip off, and lay them on the griddle pan (or at least as many as will fit spaciously, you may have to cook in batches) and cook for a minute. Using a spatula, turn the cheese slices round carefully by 90° and cook for a further minute. Do this with care as the cheese likes to test the griddle cook and will attempt to leave its desirable griddle markings attached to the pan. Again using a spatula, lift the pieces of cheese and turn them over to cook the other sides for a further minute or two. Lift from the grill and serve while still hot (with the nicely cross-hatched markings uppermost, of course).

A note on cheeseboards and not cooking with cheese

It's fairly obvious quite how much I love cheese. I think perhaps it is the love and craft of the cheesemaker that makes it such a unique foodstuff. Certain cheeses respond well to a cooking process and fit in well to my regime. A salad of grilled goats' cheese that has some French beans scattered through it and a generous sprinkling of toasted pine nuts (low, low-carb) will always fit the bill.

On my regime, cheese on toast had to go out the window (see page 104), but a plate of melted cheese with an appropriate garnish is a fabulous comfort, as in the Baked Cooleney wrapped in bacon on page 102. You can also take an Alpine cheese like raclette, slice it thinly and then let it liquefy under the grill or even in the microwave and garnish it with a mini

salad of cornichons and sliced shallot. Do be careful, though, not to allow it to colour under the grill as this will ruin the flavour.

Of course, the best thing of all is to simply eat cheese at its optimum moment of maturity and – very importantly – at room temperature, with a leaf salad to the side and a good bottle of red wine. Good cheese is expensive, but really worth it. The best thing to do when putting together a cheeseboard is to get a range of those cheeses that are in season at the time, like goats' cheeses in summer and mountain cheeses like Vacherin in autumn/winter.

GREEK SALAD

I could live on this salad in the summer – and on annual holidays in Crete I do. It is so very simple – but you must follow the rules: never use anything other than a decent Greek feta; only use tomatoes with flavour; and don't leave out that tiny sprinkling of oregano. It is the simplicity of the dressing that makes this dish – it is a simple vinaigrette, that is to say a good bathing of olive oil and vinegar. Yes, it is messy and juicy, and in the old days I would use large quantities of bread to mop up the exuded liquid, now I throw my passable table manners to the wind and drink them straight from the plate or bowl (occasionally using a spoon).

Serves 4

1 medium cucumber
8 of the ripest tomatoes
1 large red onion
200g real Greek feta

1 tsp dried oregano
extra-virgin olive oil
red wine vinegar

Peel the cucumber and cut it into thick slices. Place these in a colander and sprinkle them generously with salt. Leave for a mere 5 minutes and then rinse very well to remove the salt and pat dry with a tea towel.

Using your sharpest serrated knife, cut your tomatoes into slices that are slightly thinner than the cucumber. Lay out the tomato slices on 4 plates as the base layer of the salad, then scatter over the cucumber slices.

Slice the onions into rings and arrange these over the cucumber. Season with pepper and a modest amount of sea salt.

Slice the feta into 4 nice chunks and place one chunk on top of each salad, then sprinkle the oregano over the cheese.

To serve, place bottles of olive oil and vinegar on the table. Each person then dresses their own salad. The vital part is to break up the feta so it is distributed through the salad. It is this act that brings everything together and, as the salad ingredients become lightly bruised, they release their juices.

You can quickly see that I make fairly wide use of vegetables in my regime, as I think it is quite essential that the system also gets all the fibre it needs while eating so much protein. Moreover, veg are just so full of such a wide range of things that are good for us, from simple essential vitamins and minerals to the extraordinary range of phytochemicals that have recently been seen to help ward off a wide range of illnesses, from cataracts and hay fever to cancers and heart disease – and most of these are not available in any commercial form of supplement.

As I explained in my introduction, I do restrict myself in some ways with respect to vegetables, avoiding the starchy and sweet roots – like potatoes, parsnips and carrots – because of their high-carb content, while really favouring greens.

Some of the dishes in this section are quite substantial main-course salads, while others are very much just vegetable side dishes and some almost like garnishes. However, don't think that it is hard to make a salad substantial without added potatoes or croutons. Boiled eggs, cheese – especially goats' cheese – and meats like the kofte on page 114 and the confit duck on page 115 – all work really well.

You will notice in the recipes in this section a distinctly North African flavour and some repetition of spicy ingredients like chillies and harrisa, as I think they give this sort of food a memorable kick that really lifts it.

As with most ingredients, try to use the vegetables and leaves that are in season, as they will not only have more flavour and nutrients but should be considerably kinder to your pocket.

ASPARAGUS, LAMB'S LETTUCE AND SMOKED SALMON SALAD WITH TARRAGON DRESSING

Green, summery and fresh... enjoy this salad from the end of April until mid-June, when English asparagus is available, then store away this recipe and don't eat it again until the following year. Sprue, which is the skinniest and cheapest grade of asparagus, is best for this dish. I love salmon eggs, though I accept they are not to everyone's taste. I do believe, though, that they are full of incredibly beneficial essential fish oils. Should lamb's lettuce be difficult to find, then use an equivalent quantity of a supermarket small leaf mix.

Serves 4 daintily or 2 more substantially

500g English asparagus (see above)
150g lamb's lettuce
2 tbsp finely chopped shallot
150g smoked salmon
100g jar of salmon eggs (optional)

for the tarragon dressing
4 tomatoes
3 tbsp red wine vinegar
½ sachet of fresh tarragon, picked
4 tbsp Kikkoman soy sauce
3 tbsp crème fraîche
pinch of cayenne pepper (or a good milling of black)

Cook the asparagus in salted boiling water for 2–3 minutes, until tender. Lift it from the pan and refresh it in a big bowl of cold water (to stop it cooking and set the colour to a more vivid green). The addition of ice to the water if you have sufficient is highly recommended. Drain and allow the asparagus to rest on a tea towel to dry.

Make the tarragon dressing: put all the ingredients in a liquidizer and blitz until smooth. Pass this mixture through a fine sieve and set aside until required.

Place the lamb's lettuce in a bowl together with the shallot. Cut the tips from the asparagus and put them with the salad. Thinly slice the remaining tender part of the asparagus stalks and add these to the salad, discarding the fibrous remains.

Add three-quarters of the dressing to the salad, mix it with a light hand and serve it up on 4 plates. Cut the smoked salmon into strips and scatter over the salad. Stir the salmon eggs into the remaining dressing, spoon this over the top of the salad and finish off with a decorative milling of black pepper.

LAMB KOFTE AND GOATS' CHEESE SALAD

Preserved lemons used to be found only in Levantine groceries, but luckily most gourmet sections of good supermarkets now carry them. Preserved lemons are a bit of an anomaly as it is the flesh that is discarded and the rind that is used, as the curing process removes all bitterness.

Serves 4

2 red onions, halved
1 large Cos lettuce
½ cucumber, cut into 4cm lengths
bunch of parsley, picked
½ bunch of mint, picked
handful of cherry tomatoes
200g aged goats' cheese (moist/crumbly)
2 preserved lemons, pulp discarded and
   rind cut into strips

for the kofte
1 tsp cumin seeds
1 tsp coriander seeds
500g lamb mince
2 garlic cloves, crushed

grated zest and juice of 1 lemon
grated zest of ¼ orange
2 tsp harissa
2 tbsp chopped fresh coriander
2 tbsp chopped fresh mint
2 scant tsp sea salt
olive oil, for frying

for the harissa dressing
1 garlic clove, crushed
1 tsp harissa
2 tbsp red wine vinegar
2 tbsp Greek yoghurt
3–5 tbsp milk
3 tbsp good olive oil

Prepare the kofte: toast the cumin and coriander seeds in a dry pan until aromatic. Grind to a powder using a pestle and mortar or spice mill. Place in a bowl with the remaining ingredients and mix well. Shape into small oval patties, cover and chill for an hour.

Prepare the onions for the salad: remove the root bases and cut the onions into the thinnest of slices from top to toe. Place in a bowl of iced water and leave for an hour. They will become more crisp and their acidity/harshness will soften.

Make the harissa dressing: combine the garlic, harissa and vinegar with a good seasoning of salt. Mix well, then stir in the yoghurt, followed by enough milk to thin it down to the consistency of single cream. Stir in the olive oil and set aside.

Discard the outer lettuce leaves and cut the lettuce into 2cm wide strips, starting at the top and cutting parallel to the base. Wash and dry well. Place in a large salad bowl.

Heat a large cast-iron pan or griddle and cook the kofte in batches, in a little hot oil until uniformly golden brown all over. Keep warm while you finish the salad.

Stand the cucumber lengths on their ends and slice vertically, then slice the resultant sheets into strips. Add to the salad bowl with the herbs, tomatoes and goats' cheese.

Pour three-quarters of the dressing over the salad and mix together with a light hand. Divide the salad between 4 plates, arrange the kofte over that and finish with a scattering of preserved lemon strips and a final drizzle of the remaining dressing.

DUCK, GINGER AND POMEGRANATE SALAD

I use confit duck legs for this salad or, from time to time, I will buy half a roast duck from a Chinese supermarket and shred that instead. Increasingly, nowadays, confit and goose and duck fat are available here, at a price.

This is a fairly meaty salad and it makes a lively starter to a meal when entertaining. I was complaining to my friend Derry that fruit was off the menu on my regime as, apart from the raspberry, there was really nothing else. 'Pomegranate' was his immediate riposte. At 11 grams of carbs per 100 grams, it should be off-limits, but as it is eaten in such small quantities it must be included.

A note for the carbohydrate-consumer; a grilled pitta or Arabic flat bread is rather good wrapped around, or eaten with, this salad. A further shopping note, it is those pomegranates sold in Persian stores around October time that are the most flavourful.

Serves 4 as a starter

2 confit duck legs
1 large ripe pomegranate
2 good bunches of spring onions
handful of walnut halves
fluffy bunch of fresh coriander, picked

for the dressing
small knob of ginger
juice of 1 small lemon
5 tbsp olive oil
1 tsp poppy seeds

Preheat the oven to 180°C/gas 4. Place the duck legs, skin side down, in a heavy ovenproof pan. Place over a moderate heat and, when it starts to sizzle, transfer the pan to the oven and bake for 20 minutes, or until hot and crisp. Remove the pan from the oven, turn the legs over and set aside.

Score the pomegranate into quarters through the skin to a depth of about 1 centimetre. Submerge the pomegranate in a large bowl of cold water. While holding it below the surface, break the fruit apart and carefully tease out the little ruby orbs of flesh from the network of pith. Lift out these red seeds and set aside.

Slice the spring onions at a long shallow angle, then place them in a bowl with the walnut halves and the coriander leaves.

Make the dressing: peel the ginger and grate it over a bowl. Mix the resultant purée/juice with the lemon juice and olive oil. Season with sea salt and pepper. Toast the poppy seeds briefly in a dry pan and then add them to the dressing.

Using 2 forks, shred the meat and skin from the warmish duck, just as it is done with crispy duck in a Cantonese restaurant. Add this duck to the bowl together with the dressing and mix well.

Divide this into nice piles on each of 4 plates. Spoon any residual dressing around the edge – purely for a presentational effect – and finally scatter over the pomegranate seeds.

GRILLED CHERRY TOMATO AND GOATS' CHEESE SALAD

Tomatoes, goats' cheese and mint are a match made in heaven. This is the taste of summer with a very Provençal feel. Choose goats' cheeses that have developed a light rind as the very fresh rindless ones are almost impossible to lift from the pan after grilling. If the addition of cream to this dish seems excessive to you, then omit it.

Serves 4

500g ripe cherry tomatoes
2 tbsp chopped shallot
Basic vinaigrette (page 28)

½ bunch of mint
100ml whipping cream
4 Crottin-type goats' cheeses with a
   light rind (see above)

Preheat the grill. Arrange the tomatoes in a single layer on a grill pan. Place under the grill as close to the heat as possible. Keep a close eye on them and when most have split from the heat and some are starting to blister and their skin blacken, take them from the grill and transfer them to a dish.

Scatter over the shallot and a goodly slug of vinaigrette. Cover and leave for 1 hour.

Pick the leaves from the mint, chop them and stir them into the whipping cream. Season well with salt and freshly ground black pepper. Spoon this mixture over the tomatoes and mix with the gentlest of actions.

Preheat the grill again. Cut the goats' cheeses in half, place them on a grill pan and grill until the surfaces have coloured slightly and they are starting to bubble.

While the cheeses are grilling, divide the salad between 4 shallow soup plates. When the cheeses are ready, carefully lift them from the grill tray and place 2 halves in the middle of each salad.

GRILLED COURGETTE AND MINT SALAD

You will have some mint left over from the Grilled Cherry Tomato and Goats' Cheese Salad and this is an ideal use for it. Serve this salad as a garnish for a roast fillet of cod or perhaps to accompany some grilled or roasted lamb.

Serves 4

6 good-sized courgettes
good slug of olive oil, plus more for
    brushing

2 tbsp chopped fresh mint
2 garlic cloves, finely chopped
splash of red wine vinegar

Preheat a ridged griddle pan or, failing that, a heavy frying pan. Cut the tops from the courgettes and slice them thinly lengthways (the lethal mandolin is ideal for this).

Brush the slices with a little olive oil and grill them briefly on each side. There can be no time defined for this stage, your eyes must be your guide. A light marking from the grill and a change in texture will tell you the courgettes are ready.

Lift them into a bowl. Mix the mint, garlic, some salt and pepper, a good slug of olive oil and a splash of the vinegar. Spoon this mixture over the courgettes and mix gently. Leave for 20 minutes, during which time the dish will lose its colour and collapse even more. Don't worry, as this is one of those dishes that is about flavour rather than appearance.

RAW FENNEL, RED ONION, THYME AND LEMON SALAD

I first served this salad with some seared scallops, and a fine combination it turned out to be. In fact, it is a wonderful accompaniment for most grilled meaty fishes. The fennel is not exactly raw as, if you make it a couple of hours before you wish to eat it, the lemon juice 'cooks' the fennel and renders it more appealing. At the risk of repeating myself, the Japanese mandolin is the ideal tool for the slicing in this recipe.

Serves 4 as a salad or 8 as an accompaniment

4 fennel bulbs
1 red onion
2 tsp picked thyme leaves
juice of 3 lemons
good glug of extra-virgin olive oil

Cut each fennel bulb in half and trim any tired edges away. Using your sharpest blade, slice the fennel as thinly as you are able. Then slice the onion into the finest of rings.

Place the fennel and onion in a bowl and mix in the thyme, lemon juice, a good seasoning of sea salt and pepper and a good glug of oil. Mix well and cover. Set aside for 2 hours before serving.

If after two hours you are not ready to use the salad, then refrigerate it, remembering to remove it from the fridge at least 30 minutes before you do want to serve it.

PIEDMONTESE PEPPERS

These are quite delicious when still slightly warm from the oven. If you want to embellish them further, then top them with slices of buffalo mozzarella after baking (or put the cheese on for the last few minutes in the oven).

8 ripe tomatoes
4 red peppers
4 garlic cloves
8 canned anchovy fillets in olive oil, drained
good olive oil

Preheat the oven to 150°C/gas 2. Blanch the tomatoes for 15 seconds in boiling water to loosen the skins, refresh in cold water and then lift off the skins. Cut the peppers in half, leaving the stalks attached, and scoop out the seeds.

Thinly slice the garlic and put some at the bottom of each pepper. Cut each tomato in half and then cram two halves into each halved pepper. Cut each anchovy fillet lengthwise and put 2 pieces on top of each pepper.

Transfer to a roasting pan and liberally douse with olive oil and a milling of pepper. Bake for 1–1½ hours or until soft and lightly scorched.

BROCCOLI AND SALTED BLACK BEANS

I adore broccoli, with one caveat – that it is cooked properly. Serve it too al dente and it lacks delicacy, as its brassica heritage will overpower.

On trips to Chinese restaurants I can never resist a plate of iron-rich greens with black bean sauce. One evening at home a near-empty fridge offered up only broccoli from the veg box – hence this dish.

Serves 4 as a simple starter or as an accompaniment for roast pork – or even a simple grilled chop

2 heads of broccoli
1 lemon, quartered

for the dressing
2 tbsp Kikkoman soy sauce
2 tbsp sesame oil
30g salted black beans
1 small red chilli, deseeded and finely chopped
½ bunch of spring onions, thinly sliced
1 sachet of fresh coriander, picked and chopped
1 knob of ginger, peeled and finely chopped
1 garlic clove, finely chopped
150ml vegetable oil

Make the dressing: place the soy sauce, sesame oil and black beans in a bowl and mash with a fork to a coarse paste. Add the remaining ingredients and mix well. Set aside.

Heat a large pan of salted water. Trim the florets of the broccoli from their stalks and cook them in the boiling salted water for 4 minutes, or until tender. Lift from the water and drain briefly in a colander.

Transfer the dressing to a small pan and warm briefly for 1 minute. Arrange the broccoli on plates, spoon over the dressing and serve with lemon wedges.

TOMATO AND PRESERVED LEMON SALAD

This salad benefits from being left to stew in its own juices. It is important to leave the skins on the tomatoes, as it helps prevent the salad breaking down into a coarse slop. Serve the salad alongside some barbecued fresh mackerel.

Serves 4

½ tsp cumin seeds
1 garlic clove, chopped
½ tsp dried oregano
150ml olive oil
3 tbsp red wine vinegar
Maldon sea salt
8 ripe tomatoes
2 red onions
2 preserved lemons (see page 114; if small, such as those sold under the most excellent Belazu label, then use 4)

Toast the cumin seeds briefly in a small frying pan. Grind them to a powder using a pestle and mortar or spice mill. In a bowl, mix this with the garlic, oregano, olive oil, vinegar and a good seasoning of sea salt. Set aside.

Slice the tomatoes thinly (a serrated knife is best used for this task) and lay them out in a shallow bowl. Slice the red onion into fine rings and scatter them over the tomatoes. Quarter the preserved lemons and cut out and discard the interior flesh. Cut the rind into thin strips and scatter these over the onions. Pour the dressing over the salad and mix it all together. Cover and leave for at least 4 hours before serving.

Add more oil and vinegar to the leftover juices to make an aromatic vinaigrette for a later salad.

GAZPACHO

This makes a great quantity and it really is not worth making in smaller batches. People will often come back for more and if you make it at the start of the weekend you will finish it by Sunday night. Well, I do! This Spanish liquid salad traditionally uses bread as a thickener, but it is not necessary. I have tried making this without the addition of ketchup and it tastes flat by comparison.

Serves 10 or so

2 cucumbers, peeled
1 red pepper, deseeded
1 green pepper, deseeded
1 onion
5 large ripe tomatoes
100ml red wine vinegar
350ml olive oil

small bunch of mint
1½ tsp sea salt
2 large garlic cloves
1 chilli, deseeded
3 tbsp Heinz tomato ketchup
450g can of chopped tomatoes

Cut all the vegetables to a manageable size and then blitz everything together in a liquidizer with 350ml water. Pass through a coarse strainer (though a good strong blender does mean you can omit the straining). Check the seasoning and chill for a couple of hours before serving.

Notes on garnish: serve it plain if you desire, but a drizzle of olive oil, some chopped parsley or even a little grated hard-boiled egg will enhance the presentation.

PUDS

Puds of all sorts are quite delicious and everyone loves them — nobody more than me — and we almost feel the need for them to round off a meal and tell the body to switch off the appetite mechanism. To pursue my regime, however, I had to give them up as they were all deadly.

I looked through low-carb diet books and diet magazines, and did indeed find many, but they all used artificial sweeteners. In desperation, I tried making a chocolate mousse using a popular brand of artificial sugar and, quite frankly, it did not taste nearly as good as using the real thing.

This drove home to me the point that my regime is as much about eating good food, proper food as it is about losing weight. To then advocate the use of artificial alternatives goes against all I believe in as a cook. There are also all the possible health issues with regard to such chemicals and I certainly wouldn't want to give such stuff to my kids.

So, in a nutshell, if you want lots of puds then that's unfortunate as you won't find many here apart from the Raspberries and Jersey cream overleaf and the two in Menus. Yet again, it is a matter of adjusting your thinking to see a sweet course as an occasional treat and not the natural conclusion to meals.

If necessary, you can always fall back on less sweet fruit, such as berries, grapefruit, pomegranates and gooseberries. A word of warning about ready-made desserts which may make claims to being sugar-free, they often use in its place concentrated apple juice or chemical sweeteners. If you find yourself yearning for that sweet hit after a meal, or at any time, go for that one little square of ultra-dark chocolate.

RASPBERRIES AND JERSEY CREAM

Every new regime requires change. For me this change in eating habits brought added benefits, one of which was the application of restraint when it came to eating. A lack of discipline meant that I never said no to food of any variety or, sadly, also of any quantity. Learning to say no to certain food groups was not as hard as I thought it would be. I gave up sugars and this meant saying goodbye to nearly all fruits. Luckily my favourite, the raspberry, proved to have the lowest carbohydrate content, so I could leave them in my diet and not feel guilty drowning them in lashings of thick cream. I have benefited from this new discipline and not just from a physical point, but also mentally. It may sound pompous but I now enjoy saying no to certain things and don't just sit there sulking, wishing I had said yes. Mind you, my eye does sometimes linger over the banned sugar shaker when I am offered raspberries.

Jersey cream and double cream have the highest fat content and therefore the lowest carbohydrate content. If you are lucky enough to be able to access a farmers' market or good dairy/cheese shop, then you may well be able to get your hands on some rich yellow untreated Jersey cream.

If that is the case, then this most simple of recipes is as follows.

For each person, allow 125g ripe – preferably late season – raspberries. Serve in bowls along with a crock of Jersey or thickest double cream, your love of cream will dictate how much you will purchase.

MENUS

Entertaining needn't really be a problem in any way when you are following this type of regime. After all, give most people a nice hearty roast with all the trimmings, and they will feel suitably honoured and happy. Of course, it is the trimmings that can be a problem, so I have occasionally added in an extra side dish for those happy carb consumers, which you can leave well alone, thank you.

The principal area of difficulty is undoubtedly the sweet course, as you can readily see in the Puds chapter, pages 124–7.

As with the trimmings, possibly the easiest thing is for you to make something just for your non-dieting guests, but I have suggested a couple of sweet dishes that you too can enjoy – both heavily reliant on low-carb spirits for their flavour, and kick.

In the same spirit

(sorry, quite inadvertent), I have finished the book with a postscript of several cocktails based on low-carb wines and spirits, so – dessert or no – you can always finish – or start – a meal with a flourish.

I have also included in this section my 'Take two chickens', an exercise in good old-fashioned home-economy for those concerned that this sort of regime is necessarily expensive. The roast chicken carcasses are used to make a stock that forms the basis for a soup and then the leftover meat is stretched into a salad.

No one seems to bother that much about this sort of thing any more. Instead, we seem to prefer to head down to the supermarket and buy something quick and convenient…. And those valuable remains from that lunch? Invariably in the bin. And why? 'Oh, making stock… that's a bit complicated'.

A favourite dinner party

This is a lovely satisfying meal that has the really winning attribute of involving the minimum of preparation and cooking when your guests are with you, so you can really enjoy their company and the evening yourself.

Jambon de bayonne, celeriac rémoulade

Côte de boeuf, sauce béarnaise

Vodka and lemon tonic jelly

Serves 8

JAMBON DE BAYONNE WITH CELERIAC RÉMOULADE

This is best made the day before you want to eat it. It is an ideal dinner party dish as it is easy to plate at the last minute.

Serves 8

1 large head of celeriac or 2 smaller ones.
16 slices of Bayonne ham, or its equivalent from Parma or even Serrano
3 tbsp capers

for the rémoulade sauce
4 egg yolks
6 canned anchovy fillets, drained
3 tbsp Dijon mustard
splash of red wine vinegar
350ml vegetable oil
tiny splash of hot water

First make the rémoulade sauce: place the egg yolks, anchovy fillets, mustard and red wine vinegar in a food processor. Process to blitz the mixture to a smooth state and then add the oil, initially drop by drop, building up to a more confident stream as the mayonnaise forms. When all the oil is incorporated, add a tiny splash of hot water. Adjust the seasoning, adding some black pepper, of course, but perhaps more mustard may be needed. It should be spiky and assertive.

Peel the skin from the celeriac. Now, you need to cut the celeriac into the finest of strips. Use one of two methods: slice the celeriac as finely as you can and then take these fine slices and cut them into spaghetti-like strips; alternatively, should you possess one of those lethal, finger-shredding Japanese vegetable mandolin graters, cut the celeriac into manageable chunks, attach the medium-strip attachment and shred away, using the safety guard to ensure you don't serve your fingers as well. Place the celeriac in a bowl and mix in the mustard mayonnaise. Hands work best here. Cover and refrigerate until needed.

To serve, place a pile of the celeriac in the middle of each plate. Arrange 2 slices of ham around the edge of each pile, then sprinkle over a few capers.

CÔTE DE BOEUF
WITH SAUCE BÉARNAISE

This is one of my favourite cuts of beef, if it is well marbled – with that wonderful vein of fat encircling the eye of the meat, which ensures that the meat bastes itself and enhances succulence. A sharp fragrant buttery condiment on the side further highlights great beef. Eat beef less regularly, but buy better quality and a côte de boeuf, being the rib of beef, should have a short rib bone attached. The method below is the easiest way of serving this dish in the average domestic kitchen. Should you have a ridged grill and enjoy voluminous clouds of savoury smoke in your kitchen, then use it; alternatively, grill them on the barbie outside. Yes, it does seem a lot of buttery Béarnaise sauce, but whenever friends come for dinner this quantity is always consumed.

Serves 8

4 ribs of beef, each about 450g
good splash of oil
2 bunches of watercress

for the sauce béarnaise
bunch of fresh tarragon or 2
   supermarket plastic packs
5 tbsp white wine vinegar
6 black peppercorns
blade of mace (if you don't have it
   don't panic)
1 bay leaf
6 egg yolks
400g unsalted butter

Take the meat from the fridge at least an hour before you wish to cook it.

Start preparing the sauce béarnaise: pick the tarragon and set the leaves to one side. Put the stalks in a small saucepan together with the white wine vinegar, black peppercorns, mace and bay leaf. Add a good splash of water, bring the liquid to the boil and then simmer until half the liquid has evaporated. Strain the vinegar reduction through a sieve into a bowl and allow to cool.

Preheat the oven to 180°C/gas 4. Season the beef very well. Put the oil in a heavy pan and put that over a highish heat. Brown the beef on both sides and transfer it to a roasting dish. Place in the oven for 8–10 minutes. Remove the meat from the oven, transfer to a warm plate and allow it to rest in a warm place for 15 minutes. (I am assuming that the home cook has a reasonable confidence to cook beef. The timing given is approximate as if the rib of beef is wide and thin it will require substantially less cooking; but if it is tight and thick it may well need more – the 8–10 minutes should give a rare and juicy result.

While the meat is resting, it is time to finish the Béarnaise sauce. The most foolproof method is to use the food processor. Whiz up the egg yolks and vinegar reduction until pale. Meanwhile melt the butter and ensure it is piping hot. Then it is a

simple matter of following the rules for mayonnaise. With the machine running, pour in the melted butter drop by drop, increasing your pouring speed once the mixture has emulsified and started to thicken. Keep a kettleful of hot water to hand and should the mixture become too thick and have the look of being about to split, then a splash of hot water will save the day.

When this is done, chop the tarragon leaves and then stir them into the sauce together with a seasoning of salt and freshly ground black pepper to taste. Should, horror of horrors, the Béarnaise sauce split, then it is fairly easy to correct: transfer the split sauce to a jug, place a splash of boiling water in the food processor and then drizzle the split sauce into it at a slow and steady speed from the jug. The sauce should then come together again nicely.

To serve, slice each rib into 6 or 8 pieces and divide it on to 2 plates along with a generous garnish of watercress. Repeat the process until all are served. The Béarnaise is best placed in a bowl on the table. As for the bones, there are always a few who like to gnaw on the crisp attachments to the bone, so let them.

VODKA AND LEMON TONIC JELLY

Yes, I know that the tonic has sugar in it (you can, of course, use diet), but when this quantity is divided among eight it isn't that much per person. This is really more of a token pick-me-up to finish the meal than an actual pud. Beware, though, it does pack a bit of a punch.

Serves 8

7 gelatine leaves
400ml premium vodka
400ml tonic water
juice of 2 lemons

Soak the gelatine leaves in cold water until they soften. Lift them from the water and melt them in a small pan over the gentlest of heats. Do not boil it or it will spoil. Stir in 100ml of the vodka until all is well mixed. Pour the remaining vodka and the tonic water and lemon juice into a bowl and stir in the gelatine mixture.

   Divide it into 8 small glasses and transfer to the fridge. Leave to set, preferably overnight. Serve with a spoon and, if you don't mind about more sugar, whip up some double cream sweetened with sugar and place a dollop on top – or, of course, a small ball of vanilla ice cream is always acceptable.

A Sunday lunch

This is one of those meals that is well suited to a homely family gathering, but could be equally well suited to occasions when you are entertaining. Again it is deceptively easy, while looking splendid, and is guaranteed to satisfy and impress.

Endive and toasted walnut salad with Roquefort dressing

Roast leg of lamb with anchovy and garlic, slow-cooked lentils and creamed leeks

Pineapple with rum and crème fraîche

Serves 6–8

ENDIVE AND TOASTED WALNUT SALAD WITH ROQUEFORT DRESSING

By endive, here I mean what is known as French or Belgian endive or witloof, confusingly called chicory by greengrocers after the name of the family to which it belongs. Skinning walnuts is a veritable pain but one that is well worth the effort as the delicacy of the walnut flesh is most satisfying.

Serves 8

2 cupfuls of walnut pieces
600ml milk
8 large heads of Belgian endive
  (chicory)
4 shallots, thinly sliced
4 tbsp chopped parsley

for the Roquefort dressing
300g Roquefort cheese
1 garlic clove, crushed
150ml red wine vinegar
dash or two of Tabasco sauce
2 tsp Dijon mustard
400ml whipping cream

Put the walnuts in a heatproof bowl. Scald the milk (i.e. heat it to just below boiling point) and pour it over the walnuts. Set aside for an hour to go cold.

In a large bowl and using your fingers, break the cheese into small pieces. Add to it the garlic, red wine vinegar, Tabasco, mustard and whipping cream, with some salt and pepper if required. Still using your fingers, gently mix these together into a lumpy liquid mass. Check the seasoning and add a little more Tabasco and some salt if required. Store in the fridge until required.

Using a small fine-pointed knife, gently prise the brown skin from the walnuts. Break the white flesh into smaller pieces.

Break the endive into leaves and place these in a bowl. Add the shallot, parsley, walnuts and the dressing. (If the vinegar has caused the dressing to become too thick, it can be loosened with a splash of hot water from the kettle.) Mix the salad well and divide it between serving plates.

ROAST LEG OF LAMB WITH ANCHOVY AND GARLIC, SLOW-COOKED LENTILS AND CREAMED LEEKS

The inspiration for this recipe came from David Lea Wilson of The Anglesey Sea Salt Company, who produces not only the finest sea salt crystals that I have ever tasted but he also blends them with a magical selection of spices that he brought back from Zanzibar to produce an aromatic alternative he calls Halen Mon Spiced Anglesey Sea Salt. Don't worry, if you can't get a hold of it, the recipe works well with plain sea salt. Spring lamb has never tasted so good.

I include the lentils as a garnish for those entertaining non-dieting guests. You don't have to eat them yourself.

1 leg of lamb, about 2.5kg, boned and rolled with shin bone attached
4 garlic cloves, cut into slivers
2 small sprigs of rosemary
4 anchovy fillets, coarsely chopped
2 tsp Halen Mon Spiced Anglesey Sea Salt (see above)
3 tbsp olive oil
1 tbsp butter
200ml dry white wine
250ml strong lamb stock

for the slow-cooked lentils
500g green lentils
2 onions
4 celery stalks, halved
4 fresh bay leaves
splash olive oil

for the creamed leeks
8 large leeks
150g unsalted butter
250ml whipping cream
1 tbsp grainy mustard
3 tbsp chopped fresh mint

Using a sharp narrow-bladed knife, cut about 15 incisions in the lamb and stuff each with a piece of garlic, a rosemary leaf or two and a piece of anchovy. Place the lamb in a large roasting pan and rub the salt generously all over the outside of the joint. Set aside in a cool place for 1 hour to allow the flavours in the salt to do their bit.

Preheat the oven to 220°C/gas 6. Drizzle the olive oil over the lamb, dot with the butter and transfer to the oven. After 20 minutes, baste the lamb, pour in the white wine and reduce the temperature to 180°C/gas 4. Cook for a further 15 minutes per 500g for a nice rosy medium. Continue to baste the meat with the surrounding liquid.

At least an hour before serving, start cooking the lentils: place all the ingredients except the oil in a large heavy pan and season with pepper only (salt hardens skins at this stage). Cover generously with water, bring to a simmer and continue to simmer as slowly as possible. Depending on the lentils, cooking time can vary from 30 minutes up to an hour or more. Keep an eye on them, as you don't want them overcooked, floury and soupy. When cooked, remove from the heat and season well

with sea salt. Lift out the vegetables and herbs, strain and stir in splash of olive oil.

At the same time as the lentils are cooking, cook the leeks: thinly slice as much of the leek as possible into rounds. Melt the butter in a heavy pan and add the leeks. Season, add a splash of water, cover and cook on the gentlest of heats for about 30 minutes, or until completely softened. Remove the lid, add the cream and cook for a further 15 minutes, or until the cream has thickened sufficiently to coat the leeks. Stir in the mustard and mint, and adjust the seasoning if necessary. Keep warm.

Remove the cooked lamb from the oven and transfer it to a warmed serving platter, then allow it to rest for 20–30 minutes in a warm place while you finish the gravy. Place the roasting pan on the hob, bring its contents to the boil and add the lamb stock. Stir well with a wooden spoon, scraping up all the sediment, to deglaze the pan, and reduce this liquor until you have a nice flavourful gravy, taking care not to make it too salty. If it is over-salty, bring it back to the boil and whisk in a couple of generous knobs of butter.

To serve, carve the lamb and serve with the lentils, leeks and gravy.

PINEAPPLE WITH RUM AND CRÈME FRAÎCHE

Here we have sugar rearing its ugly head again – no one will force you to eat it though, apart from yourself. What I like about this dish is the ease of serving and, with the bottle of rum on the table for each diner to self-administer from it, those who don't want rum can leave it alone.

I know it's the second pud laden with booze in this book… I think it's because, when I was a child, most of my mother's puds were heavily seasoned with alcohol. Take her brandy sauce recipe at Christmas – 'Make two pints of crème brûlée mixture and add enough Armagnac to render it pourable!'

Serves 8

2 ripe pineapples
500g crème fraîche
bottle of dark rum, such as Planters or
   something nice from Martinique

Top and tail the pineapple and then carve off its thick skin. Slice the flesh thinly and arrange on a serving platter. Transfer the cream to a bowl and serve alongside the pineapple and the bottle of rum.

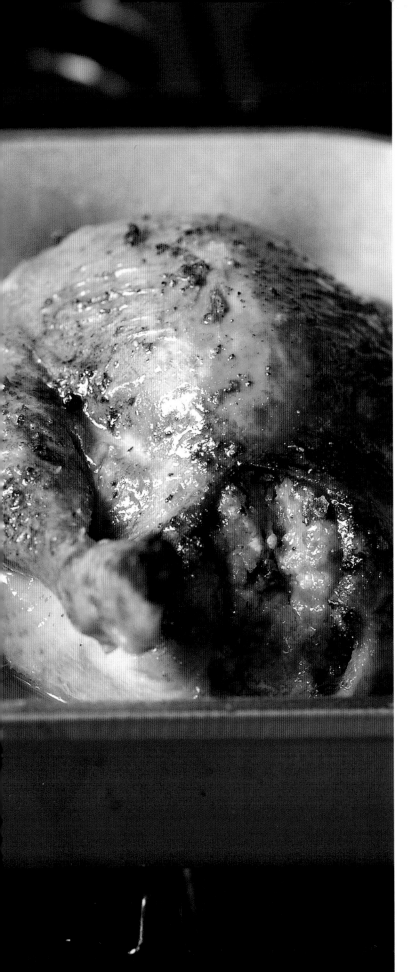

Take two chickens

As a child, a Sunday roast meant all sorts of things, but one of the less important points as far as I was concerned — though very important for the domestic cook — was that it provided several more meals. Cold cuts provided sandwiches or a salad on Monday and the carcasses made a stock that formed a basis for a soup, etc.

Roast chicken with a thyme, parsley and lemon stuffing

Straightforward chicken stock

Egg and lemon soup with oregano

Chicken, prawn, celery and almond salad

ROAST CHICKEN WITH A THYME, PARSLEY AND LEMON STUFFING

With all that pork, you are going to get the best gravy. Leftover stuffing? Slice it and refry it for breakfast, either in place of or alongside the bacon.

Serves 6–8

2 good-quality free-range chickens,
    each about 1.5kg
200g butter, softened
500ml white wine

for the thyme, parsley and lemon stuffing
1kg pure pork sausagemeat
grated zest and juice of 2 lemons
large bunch of parsley, picked and
    chopped
2 tsp picked thyme leaves

Preheat the oven to 180°C/gas 4. Make the stuffing by mixing all the ingredients together with seasoning to taste.

Remove all trussing string, etc., from the birds, as well as those nuggets of fat always left in the chest cavities. Shove half the stuffing into the cavity of each chicken, as far as it will go. Place the birds in a large roasting pan and slather them with butter, then season again. Pour the wine and 300ml water into the roasting pan.

Roast for 1½ hours, basting every 15 minutes for a crisp skin and succulent bird. A good ruse for keeping the chicken nice and moist is to cover the top loosely with foil for the first half of the cooking time (removing it to baste, of course). In my (pretty efficient) oven, the chicken is ready after 1½ hours. If nervous, use a temperature probe, inserting the tip so it hits the middle of the stuffing. If the core temperature is 80°C, then you are OK. If the liquid reduces away too quickly during cooking, top it up with some water – you want to end up with about 500ml.

Remove the birds from the oven, lift on to a large warmed platter and keep warm. Skim off excess of fat on the surface of the roasting juices, then boil them up, stirring in all those lovely brown splatters stuck to the edge of the pan. Taste and, if too salty, dilute with a splash of water.

To serve, carve slices from the breast and legs and arrange on warmed plates. Hoick out the stuffing and serve that alongside the meat. Spoon over the roasting juices (they are just that, rather than a gravy, as they are thin and flavourful, tasting of what was in the pan rather than a thickened liquid enhanced by other additions).

Step two, A GOOD STRONG CHICKEN STOCK, and if you have a pressure cooker hidden away somewhere, it just became ridiculously easy. Remove leftover stuffing and go over the carcasses, stripping every piece of meat from every bone – it's amazing how much comes off.

Put the carcasses in the pressure cooker (if you don't have one, then just simmer the stock for 4 hours instead) with any leftover gravy, 2 large onions, quartered (skin and all), a handful of parsley stalks, 2 bay leaves, 12 black peppercorns, 4 celery stalks, cut in half, a bit of white or red wine (left at the bottom of a bottle) if you have it and 1 crumbled good-quality stock cube (don't be rude about this as it seasons the stock beautifully) and add a minimum of 2 litres of water, so everything is covered.

Clamp on the lid, set the pressure to its maximum setting and place it on the stove. Bring up to temperature and, when it starts to hiss, turn the heat down a little, though make sure it continues to hiss. Cook for 45 minutes. Remove from the heat and leave for 2 hours to cool.

Strain the stock into a large bowl, discarding the solids, and measure the stock to get an idea of its volume as you now need to pour it into a pan and boil it to reduce it down to 1 litre. Allow to cool and store in the fridge. This stock will be strong, flavourful and will set to a jelly. Most impressive. Reduce the stock even more and freeze it if you plan to use it, suitably diluted if necessary, at a later date.

EGG AND LEMON SOUP WITH OREGANO

Step three, a nourishing and restorative soup. As I no longer put potato in soups, sometimes something is needed as a thickener. Egg yolks do this exceptionally well, as they impart a voluptuous quality that can't be found elsewhere. I love the flavour of oregano in both its dried and fresh forms. Sadly, supermarkets don't always carry it. If you fail in your search, then tarragon or thyme will make lovely alternatives. Using the liquidizer for two jobs may seem a pain, but it is worth it.

Serves 4

100g butter
3 large onions, halved and thinly sliced
½ bunch of celery, thinly sliced
1 small sachet of fresh oregano
4 garlic cloves, bashed

1 litre fresh chicken stock
200ml double cream
grated zest and juice of 2 lemons
yolks of 5 large eggs

Melt the butter in a suitably sized non-reactive pan and add the sliced onions and celery, season with salt and pepper, and cook very gently for about 30 minutes, or until the vegetables are meltingly soft.

Add the oregano and garlic, and cook for a further 10 minutes. Add the stock, bring it up to a gentle simmer and cook for a further 15 minutes. Remove from the heat.

Liquidize the soup in batches and pass it through a sieve, then return it to the pan. Add the cream and heat through again. Adjust the seasoning, if necessary. Place the lemon juice and zest in the liquidizer together with the egg yolks. Blitz for a minute until a good frothy mixture is arrived at.

The final step requires a little caution and attention. First, place about a third of the simmering soup into a pouring jug, pour it most slowly into the liquidizer while the motor is going at full speed and run for 1 minute. Your lid on the liquidizer should have a gap to enable you to do this safely and with the minimum of mess. Finally return this thickening mixture to the pan and, over a gentle heat, cook it for a further 2 minutes, while stirring continuously. DO NOT LET IT BOIL. Ladle into bowls.

CHICKEN, PRAWN, CELERY AND ALMOND SALAD

Step four, leftover chicken, shredded into a quick crisp salad dressed with a jar of good-quality bought mayonnaise. This is a bit of a Waldorf salad without the apples and raisins. As there is actually never quite enough leftover chicken, I often defrost 200g of frozen peeled cooked prawns and add them. Yes, it does all seem a bit quaint and straight from the pages of the Cordon Bleu Cookery magazines that my mother subscribed to in the '70s, but so what?

Serves 2–4

125g flaked almonds
1 bunch of celery
1 small red onion, halved and thinly
 sliced
300g shredded cold roast chicken
200g peeled cooked prawns

1 ripe avocado, halved, stoned, peeled
 and diced
250g mayonnaise
Tabasco sauce
2 tbsp finely chopped chives

Preheat the oven to 180°C/gas 4. Spread the flaked almonds out on a baking sheet and bake them for 10 minutes or until golden. Remove them from the oven, tip them on to a plate and leave them to go cold.

Run a vegetable peeler down the outside of the celery stalks to remove the stringy part – a bore, I know, but worth it. Then thinly slice the stalks at an angle to give nice little crescents. Put in a salad bowl with the onion.

Tip in the chicken, prawns, avocado, almonds, mayonnaise and a good seasoning of Tabasco. Taste and adjust the seasoning. Serve in piles, scattered with chives.

COCKTAILS

CLASSIC MARTINI
AND GIBSON

I learnt a long time ago that the simplest dishes are the hardest to prepare well. A fault at any stage will show up in the finished dish, whether it be poor shopping or lack of technique or skill. It is my love of food and cooking that goes a long way to rendering this first sentence near obsolete. If you love and enjoy what you do then this will stand out.

Take the Martini cocktail, undoubtedly the best drink to start an evening and – on occasion – lunch... simple, austere ingredients mixed briefly and with care. It is as much about ritual as anything else.

Gin should always be the first choice, though I must confess that its botanical properties, while fragrant and alluring, do my mood and head no good and, as a consequence, vodka is now my base of choice. If on the rare occasion that I will drink a gin-based Martini, then I like to take it as a Gibson, as I prefer the salty tang and sharpness of the cocktail onion. I acknowledge that the twist of lemon zest – and its contained oil it contributes to the drink - is to be much commended and must be mentioned to ensure that fairness is observed, but I know what I like.

Proceed as follows:

Very cold ice
A large mixing glass (I use a straight pint glass), a long spoon and a strainer
Store a bottle of Noilly Prat dry vermouth in the fridge

Store a bottle of Tanqueray gin in the freezer
Store martini glasses in the fridge
Store a jar of small cocktail onions in the fridge

Take some ice and chill down your mixing glass, discard the ice and fill it with fresh ice. Pour two capfuls of the dry vermouth over the ice, stir three revolutions and then strain out the vermouth. Pour in the gin, just over a double measure (minimum 50ml), stir three times and strain into a well-chilled martini glass. The briefness of the stirring is all-important, as over-stirring will encourage the ice to melt more rapidly and therefore water down your cocktail.

I mentioned earlier that the vodka-based martini is my preferred option and to build a martini the vodka way I like to use Ketel One vodka. Most vodkas seem to have a complete absence of taste, but it is the smooth finish of this particular vodka I find well suited to my palate.

BOURBON SOUR

Bourbon is warm, smoky and comforting, and its slight vanilla flavour makes it an appealing drink to finish a meal. A 'sour' is, in essence, a spirit mixed with lemon juice rendered palatable by the addition of a little sugar and a little egg white for a sort of frothiness. One teaspoonful of sugar is all that is required for this drink and with a carbohydrate content of 5 grams you really won't be in too much trouble.

Spirits have great digestive qualities when taken at the end of a meal, and the acidity of lemon juice is a wonderful ingredient to cleanse the palate.

I was once served this drink with a rasp of nutmeg grated across the top and have adopted this garnish ever since.

My business partner, Eric Garnier is the most elegant of cocktail makers and he produces the most excellent sour. For his version, leave out the nutmeg and instead add a few drops of Angostura bitters to the shaker. As a garnish, he drops a slice of maraschino cherry into the finished cocktail.

Note: This drink contains some raw egg white, THOSE WHO ARE PREGNANT OR ELDERLY, OR OTHERWISE WITH IMPAIRED IMMUNITY, SHOULD THEREFORE AVOID THIS DRINK.

1 tsp unrefined caster sugar
3 tbsp strained fresh lemon juice
handful of ice

5 tbsp bourbon, Maker's Mark for
  preference
½ egg white
freshly grated nutmeg

Place the sugar and lemon juice into a metal cocktail shaker and stir well to dissolve the sugar. Add the ice, bourbon and egg white, clamp on the lid and shake hard and briefly in an extrovert manner. Strain into a suitable glass. Take a small grater and give one rub of nutmeg across its surface while holding it over the surface of the cocktail. Sip.

A CHAMPAGNE COCKTAIL

The trouble with this sort of diet is that you have to say farewell to puds.
Without – I hope – sounding like an old soak, I have found solace in cocktails.
Being, in the main, spirit-based, there is a happy absence of carbohydrate.
Where some liqueurs are needed, their sugar content in the grand scheme of
things is minimal enough not to cause undue damage/weight gain.
  I was not going to offer up this recipe until recently when, at a do hosted by
Laurent Perrier, they offered their delicious Ultra Brut Champagne. What makes
this Champagne so dry, I was told, is that, unlike all the others, it has no
'dosage' – the top-up in the bottling process which contains sugar. As a result,
their Ultra Brut contains half the carbs of other Champagnes. Fab I thought.

Serves 1

Take a Champagne glass and pour in 1 level tablespoon of Armagnac or Cognac. Add 1
level tablespoon of an orange-based liqueur. Fill with Champagne, pare off a small
piece of orange zest and twist it over the surface of the Champagne to release its oil
and then drop it in. If serving this instead of pudding, then drop in a few raspberries.

INDEX

ACKNOWLEDGEMENTS

This book has turned out to be more personal than I ever anticipated. It reflects changes I have made to the way I live and I say that unequivocally. To reach this stage I could not have done it alone.

The greatest thanks go to my wife Denise. She showed me how easy it was to make the necessary changes and, even more importantly, how to maintain them.

Thanks also to:
Lewis and Gaye, my parents, for their support and enthusiasm for great food.
My agent, Jacqueline Korn, I required a little convincing and you did it so well, I never really noticed.
My editor, Lewis Esson, who removed any concerns and stresses along the way.
Jason Lowe for his photographs that say it all about real food.
Lawrence Morton, the book's designer, who brought everything together.
A special thank you for Sunil Vijayakar, food stylist. You took the recipes and cooked them and they came out just as I imagined, a rare gift.

Elfreda Pownall of *The Sunday Telegraph*. She commissioned twelve recipes from me last year for a high-protein/low-carbohydrate regime and it all set the ball rolling.
Eric Garnier, my business partner at Racine, your support and encouragement for this project is greatly appreciated.
Chris Handley and the brigade at Racine for their invaluable feedback, recipe suggestions and support.
Finally, to Alison Cathie, Jane O'Shea and the team at Quadrille. Thank you for pursuing me and being so supportive and complimentary along the way.